THE UNTOLD STORY OF CAPTAIN JAMES COOK RN

THE UNTOLD STORY OF CAPTAIN JAMES COOK RN

REVELATIONS OF A HISTORICAL RESEARCHER

COLIN WATERS

AN IMPRINT OF PEN & SWORD BOOKS LTD.
YORKSHIRE – PHILADELPHIA

First published in Great Britain in 2023 by
PEN AND SWORD HISTORY
An imprint of
Pen & Sword Books Ltd
Yorkshire – Philadelphia

ISBN 978 1 39905 696 0

A CIP catalogue record for this book is available from the British Library.

Typeset in Times New Roman 12/16 by
SJmagic DESIGN SERVICES, India.
Printed and bound in the UK by CPI Group (UK) Ltd.

Pen & Sword Books Limited incorporates the imprints of Atlas, Archaeology, Aviation, Discovery, Family History, Fiction, History, Maritime, Military, Military Classics, Politics, Select, Transport, True Crime, Air World, Frontline Publishing, Leo Cooper, Remember When, Seaforth Publishing, The Praetorian Press, Wharncliffe Local History, Wharncliffe Transport, Wharncliffe True Crime and White Owl.

For a complete list of Pen & Sword titles please contact
PEN & SWORD BOOKS LIMITED
George House, Units 12 & 13, Beevor Street, Off Pontefract Road,
Barnsley, South Yorkshire, S71 1HN, England
E-mail: enquiries@pen-and-sword.co.uk
Website: www.pen-and-sword.co.uk

or

PEN AND SWORD BOOKS
1950 Lawrence Rd, Havertown, PA 19083, USA
E-mail: uspen-and-sword@casematepublishers.com
Website: www.penandswordbooks.com

Contents

Introduction & Acknowledgements

This book is the end result of a lifelong personal quest which has taken over forty years to research and compile. It has its roots in a fascinating tale repeatedly told to me when I was a boy by older family members. My curiosity was increased when I listened to the gossip of old men sitting on benches at the harbourside in my native Whitby, North Yorkshire, as they casually discussed the life of Captain James Cook R.N. and his son's supposed desertion from the Royal Navy.

As an adult, my extended research has involved many years of scouring old books, perusing internet sites and visiting or corresponding with record offices, organisations and individuals across the world.

Thanks are due to the following institutions and experts, especially to the people named in brackets for their often enthusiastic response to my enquiries. My thanks also goes to interested correspondents, too numerous to name individually, who have supplied me over the years with tiny snippets of information, many of which unexpectedly lead to further extended research.

Bibliothèque et Archives de Canada- Library and Archives of Canada. (Gaya Déry)

Cambridgeshire Archives (Will Fenton and Yaye Tang)

Cambridge Archaeological Unit -University of Cambridge. (Samantha Smith p.p. Director))

Captain Cook Society (Cliff Thornton)

Christ's College Cambridge (Professor Geoffrey Thorndike Martin, Hon. Archivist)

Dorset County Archives, History Centre (L Garratt)
Hawaiian Historical Society (Barbara E. Dunn – Administrative Director)
Isle of Wight County Record Office (Richard Smout – Heritage Service Manager)
Isle of Wight Heritage Service (Sheila Caws – Island Heritage Librarian)
Isle of Wight Branch of the Historical Association, (Terence J. Blunden Hon. Secretary)
Isle of Wight – Ventnor L.H.S. & Heritage Museum, (Fay Brown Secretary/Researcher)
Library & Museum of Freemasonry (Peter Aitkenhead, Assistant Librarian.)
Marshall, Barbara (Cook Researcher, Australia)
National Archives of Australia (various staff)
National Archives U.K. (Various staff)
National Library of Australia (Ralph Sanderson)
National Library of New Zealand – Alexander Turnbull Library (Helen Smith and Jill Goodwin)
National Maritime Museum, Royal Museums, Greenwich (Katherine Weston and Martin Salmon)
National Museum of Australia (Michelle Hetherington – Curator)
National Museum of the Royal Navy (Mrs Heather Johnson)
National Meteorological Archive, Met Office (Glyn Hughes Assistant Archivist)
Naval Record Society (Robin Brodhurst – Hon. Secretary)
New South Wales State Library (Julie Sweeten, Librarian, Access and Information)
North Yorkshire County Record Office (Tom Richardson – Record Assistant)
Redcar & Cleveland Borough Council (Mrs T. Wills -Senior Neighbourhood Assistant Librarian)
Royal Archives, Windsor Castle (Pamela M. Clark – Senior registrar/ archivist)

Royal Navy Club, (Captain John Hall. Secretary)
Stephen Scott-Fawcett MA(Cantab) - Journal Editor - James Caird Society
St Andrew the Great Church Office, Cambridge (Amy Cooper, Administrator)
West, Dr. Ian. (Scientist – Geology of the Wessex Coast Website.)

Overview

POOLE 26 JANUARY 1794

"Capt Cook, of his Majesty's ship, Spitfire, with six of the crew were unfortunately drowned on Friday evening last, in attempting to reach the ship, which lay in anchor at Poole bar. The Captain had that day taken a farewell dinner with some of his friends at Poole, and was remarkably serious and melancholy: he left his company immediately after tea, being intent on sleeping on board, as the ship was under sailing orders – It blew fresh from S. W. and was very dark; he was advised to desist from attempting a passage that night; --- but his sense of duty prompted him to persevere!!! We are all melancholy on the occasion as he was most deservedly beloved and respected here. His body was taken up yesterday, near Christ church."

This centuries old press cutting reporting a captain's drowning at first glance appears un-noteworthy; except perhaps for the fact that it contains the familiar name of Captain Cook. This isn't of course the world renowned circumnavigator who claimed Australia for Britain and was later murdered by natives in Hawaii, but his eldest son, also named James, who, like his father, was a respected Royal Navy commander.

The death of the largely undocumented James Cook junior hides a mystery that has never before been definitively answered despite speculation by numerous historians, writers and researchers over many years.

From the day that the tragedy occurred, an official cover-up of the circumstances of the incident seems to have been rigorously put in place. Reports of the drowning changed constantly to fit an official narrative and a full Admiralty enquiry into the matter appears never to have reached a satisfactory conclusion.

Instead, James Cook junior's name seems to have been virtually expunged from all official records, even to the extent that the log of his ship, the Spitfire, contains no reference whatsoever to its missing captain, despite him supposedly drowning on his way to resume his command of the vessel on that fateful night.

Though history simply records James Cook's unfortunate drowning as a given fact, as we shall see, an alternative and very different version of events has survived in rural North Yorkshire. Local families have handed down stories insisting that the young James Cook, despite his success and social standing, became so disgruntled with naval life that he actually deserted his post in order to return to North Yorkshire and be reunited with his son and the boy's mother. This story gives credence to people throughout the world who have over the years claimed direct descent from the famous Captain James Cook senior.

Luckily one substantial set of documents concerning the mysterious drowning incident *has* survived. These remarkable handwritten official papers have laid largely unseen and gathering dust for well over 200 years in the British National Archives. When examined, they definitively answer many of the general questions that have perplexed researchers who have sought the truth behind the story of James Cook junior's mysterious final boat journey. The old documents, contained within a single file, unveil a detailed record of the Surveyors General of Customs official inquiry (followed by an immediate subsequent enquiry). They contain multiple witness statements from named individuals. Their testimonies provide a rare and invaluable insight not only into their personal details (e.g. age, job and literacy), but also into the general working practices, attitudes and even the personalities of mariners serving aboard 18th century naval Coastguard cutters.

Despite these mariners giving evidence under oath, it's clear that they could not, or would not, agree on exactly what happened. It was claimed by some sailors that the unpopular Stephen Watson, commander of the Excise cutter Greyhound was solely responsible for the death of Cook and his rowing boat crew. The revelations at the enquiry probably account for the lack of substantial public information being released at the time and could explain why the incident may have been purposely covered up to prevent a public scandal.

This book has taken over forty years of intensive investigation to compile. Its findings would seem to discount the age old premise that Captain James Cook, the great circumnavigator, has no surviving descendants. In addition, it unravels the mystery surrounding the tragic drowning and sheds new light on the exact circumstances of the incident while including a full transcript of the inquiry into the culpability of Captain Stephen Watson; the man accused of causing the deaths of Cook and his fellow mariners.

Perhaps more importantly, for the first time the findings presented here draw together genealogical information linked to the North Yorkshire rumours that Captain Cook junior fathered at least one child. Coupled with compelling evidence that he may well have staged his own death before adopting a simple farmer's life in North Yorkshire, the story uncovers the fascinating threads of an explosive history-changing mystery.

All details contained in this investigation have been verified with official newspaper reports from the time and have been clarified where necessary through correspondence with verifiable experts in academic and naval archives, museums and other wide ranging reliable authorities around the world. These include: local heritage collections, museums, national libraries and even the British Royal Archives. All major sources are listed in the acknowledgements.

Because the story of James Cook junior's death/survival is a convoluted and sometimes confusing one, this book has been divided into a number of thematic chapters and subheadings so that each part of the fascinating story can, if required, be examined in isolation.

A Convenient Death?

The name of Captain James Cook R.N. (1728-1779) is known throughout the world. During his fifty one years he became one of the world's most celebrated explorers, discovering and charting many of the totally unknown parts of the globe and placing the country of Australia on the map forever.

Conventional history tells us that Cook has no surviving descendants. More to the point, any individual who has claimed over the years to be his direct descendant has been at best discredited, or at worse ridiculed. This is usually based on the premise that none of the explorer's children survived into adulthood, an assertion that is seriously flawed.

Interestingly, one of the claimants to direct descendancy was none other than Frank Wild, second in command of Sir Ernest Shackleton's historic South Pole Expedition. The renowned polar explorer Shackleton himself confirmed and supported the claims of a Cook family descendancy using his own research.

As we know, Captain James Cook did in fact have a son who grew into adulthood. Like his father he was also named James Cook and was destined to follow in his footsteps to become a respected Royal Navy commander. More interestingly, as the evidence will show, James Cook junior's death may have involved a cover up or conspiracy at the highest levels. In such a class conscious era, the reasons for this may well have been to not only protect the reputation of James junior's respected father and mother but also to hide the fact that an esteemed Naval Commander had opportunistically deserted his post; arguably because of his intense dislike of the career his parents had mapped out for him since birth. According to local rumour, James junior returned to his father's native North Yorkshire to follow in the footsteps of his grandfather, who also named James Cook (1693–1779), a former Scottish farm labourer from Ednam in Roxburghshire, who had moved south to work as a Yorkshire farmer at Marton.

The official account has always supported the initial supposition that a body washed ashore on a beach in southern England was that of the thirty year old James Cook junior, but as we shall see, despite an official enquiry, it appears that no positive identification of the corpse was ever proven. There is however evidence that in the scandal-averse culture of the 18th century, there had been efforts to suppress the fact that the young commander had been married to a 'low class Yorkshire woman' and fathered at least one child, a boy; much to the displeasure of Elizabeth, James Cook junior's mother. Further evidence has been uncovered that James Cook junior later visited the child to assure him of the legitimacy of his birth.

Modern history books generally accept that the young naval officer drowned on the night of Friday, 24 January 1794 in an open boat whilst attempting to sail back to his ship the Spitfire during stormy weather. However, as in all great historical mysteries, there appears to be more to the story than first meets the eye. Accounts of the incident (even in newspaper reports of the time) differ widely and when one examines the circumstances more closely, a variety of puzzling questions arise.

Why for instance does there appear to be no surviving records of the young naval officer's death in prominent archives, or handed down stories on the Isle of Wight where his body was supposedly found. More intriguingly, why has his naval career been virtually obliterated from official records? Another mystifying fact with regards to the supposed drowning is that James's mother Elizabeth Cook chose to destroy everything relating to her eldest son soon after his controversial death.

Perhaps the strangest mystery of all surrounds the once persistent rumours in Sheriff Hutton and a wider area of North Yorkshire that James Cook junior didn't actually drown, as history would have us believe. Instead, it is said, he used the incident as a cover to desert from the Navy and to return to North Yorkshire where he lived out the rest of his life close to his young son and former wife. If this is true, then contrary to accepted historical accounts, many descendants of Captain James Cook (senior) the renowned explorer and circumnavigator may still be alive throughout the world today!

THE STORY BEGINS

Captain James Cook (junior)

As we have seen, James is the little documented thirty year old son of the more famous Captain James Cook RN. He was born on 13 October 1763 at his parents' home in Shadwell, London and later baptised at St Paul's, Shadwell, a place of worship known colloquially as the Church of Sea Captains. Little is known of his early life but in May 1782 James Cook Junior passed his examination for Lieutenant and was later appointed Commander of the 16 gun sloop Spitfire, replacing Commander Philip Charles Durham who had stepped down in June 1793. It was this ship that James Cook junior was sailing or rowing to, in an open boat in January 1794 when it was at anchor close to Poole Harbour off the Isle of Wight, but Cook and his men never reached it. We are told that his body was recovered and shortly afterwards buried at St-Andrew-the-Great, Cambridge.

Captain Cook junior's ship HMS Spitfire was a Tisiphone-class fire ship of the Royal Navy. Though fire ships were originally naval vessels filled with combustibles which were set on fire before being sailed into the midst of enemy fleets, by the late 1700s some were simply fitted with rockets, canons and other fire-generating weapons. The Spitfire had served as a peacetime vessel following the end of the American War of Independence but with the outbreak of the French Revolutionary Wars, she had been reclassified as a 14-gun sloop-of-war and spent most of her time in home waters between sailing to British stations in North America and West Africa.

We are told that James was of the 'Gentleman Class' of Naval commanders and as such must have been quite well off, especially after receiving a substantial inheritance as the sole surviving child of his famous father whose will had made provision for his wife and (when written) for all of his children. We should remember that if James junior's death was indeed faked, he would likely have forfeited the remainder of his impressive inheritance, which would presumably have passed to his mother as directed in Captain Cook senior's will:-

> *"...I do hereby declare that if any or either of my said Children shall happen to die before his, her or their portion or portions shall become payable, then the portion or portions of him, her or them so dying, or so much thereof as shall remain unapplied for the purpose aforesaid, shall go to the survivor and survivors of them..."*

It would appear that young Cook was well liked in naval circles. Following his 'death' newspapers immediately described him as *'an experienced officer, a loyal subject, an active friend and an agreeable companion who was without enemies...for it was impossible he could ever make one.'*

We know from a letter written by Elizabeth Honeychurch of Mile End to a Mrs McAllister in 1792 that James was living in the style of a gentleman and kept a horse. Could it be that he also held land; and if so where?

The British History online website (2015) refers to one possible holding that deserves further research, namely the largely undocumented manor of Balsdean in the parish of Rottingdean in Sussex. Balsdean is no more but was once a working village with a manor house, some farms and later a lunatic asylum before it was taken over in 1939 by the Ministry of Defence and used for military firing practice. This manor remained in the Beard family until 1792, when Kitty Beard, James Cook the younger and Mary, his wife conveyed it to Francis Whitfield. It's known that the John Beard who purchased it from a gentleman named Captain Charles Geere, took possession of Balsdean Farm in 1745, and advertised it for sale in 1790, approximately eleven years after James Cook senior died and four years before the naval commander's son allegedly drowned. At that time it was described as consisting of *'1007 acres – also 303 acres with neat farm house recently built'*.

Though purely circumstantial, until proved otherwise by more research, it would be interesting to know if this James Cook the Younger with a wife named Mary, was one and the same person as James the son of Captain Cook, the circumnavigator. It is true that the terms 'the younger' and 'junior' are used to define father and son in many family relationship but in this context it seems that there appeared to be no need to clarify which Cook family was being spoken of at a time when both Captain Cook RN and his son were so well known to everyone. Interestingly we also find James and wife Mary mentioned in legal documents in 1792.

> *"Francis Whitfeld, plaintiff, and James Cook the younger, gent., and Mary his wife, and Kitty Beard, deforciants Manor of Balsden alias Balsdean alias Ballesden alias Ballsden and tenements in Rottingdean, also all tithes whatsoever yearly arising, growing or renewing in Rottingdean, quitclaimed to plaintiff and heirs*"

The story of James Cook junior's supposed death begins on a stormy night in January 1794 when he set off from Poole harbour with a number of sailors to join their ship HMS Spitfire. Modern accounts say that he had just been appointed as commander of the ship and was rowing out on the night of his death to take up his new position. Documentary evidence, however, shows quite clearly that he was already in charge of the vessel and had been since June 1793. The London Star of 28 June 1793 reported that the previous day *'Lieutenant Cook, son of the celebrated circumnavigator is made Master and Commander into the Spitfire'*. The Sunday Reformer and Universal Register for 30 June 1793 confirmed the fact that only four days previously, the Admiralty, who had put the Spitfire into commission, had given the command of the vessel to *'Captain Cooke'*. Furthermore, Lloyd's Evening Post (London) reported that the *'Spitfire sloop of sixteen guns'* had sailed from Portsmouth *'on a cruise'* under the command of Captain Cook on 15 October 1793. This so-called cruise appears to have been a specific naval manoeuvre to meet up with forces in the war against the French. To support this theory, the London based newspaper St James Chronicle, also known as the British Evening Post, carried a report of British involvement near Tour. James Cook is reported to have brought letters to the Admiralty from Lords Hood and Mulgrave containing reports of combined army activities. It also confirmed Captain Cook had left Toulon on 30 October 1793.

As we shall see, newspaper records of Captain Cook junior's death were far from consistent. It would seem that all press reports at that time were simply variations of a standard press-release issued by the naval authorities, possibly whilst they gathered information together and decided 'how to play' the situation. Strangely, no other records appear to exist to substantiate his death at all, other than a record of the enquiry and a supposed burial service which itself may well have been 'stage-managed' to fit the story.

Though this may seem a far-fetched suggestion, the internment appears to have been hurriedly arranged without any newspaper reports listing prominent mourners or other details normally found regarding the funerals of prominent people at that time. Though the burial could of course have been a genuine one, there is still the possibility that the coffin may have contained a man who was misidentified as James Cook, as we shall examine later. Conspiracy theorists have even put forward the argument that the weighted coffin was empty and that Mrs Cook went along with the funeral either out of ignorance, or in connivance with the authorities due to the scandal that would have arisen had it become public knowledge that her well-known son had deserted his post. What is certain is that all the newspaper reports about the incident in the subsequent few days follow the same general story, though some differed widely in the material facts they describe.

At the time it would appear that nobody was sure exactly where the body identified as Cook's was really found; though most agreed that it was on the Isle of Wight, either *'on the beach near Warden Lodge'*; *'near Little Yarmouth'*; or *'close to a mill'*. Others were under the impression that the corpse was discovered in Poole harbour or that it was actually washed up on the southern mainland *'near Christchurch'*

The Times newspaper of 29 January 1794 reported that Cook junior had eaten a farewell dinner with some of his officer friends in Poole on that fateful night. Other newspapers described Cook as being 'melancholy'; and one newspaper refers to a letter that Cook had sent to a mysterious woman just before his departure.

THE MYSTERIOUS FARMER WHITE

Warden Ledge is a rocky formation that separates Colwell and Totland Bays on the Isle of Wight. Young Captain Cook's body was said to have been recovered by a Mr White, a farmer from South Yarmouth on the island. Oddly, records don't record a farmer (or miller) named White living anywhere near the spot, even though customs officers from Cowes who went to the area to investigate the case reported that they had interviewed a 'sheep farmer' who had discovered the body. White was never actually mentioned by name in any later report or in official documents presented to the inquiry. Could it be that the farmer was in fact an invented character in an elaborate cover up, whose name was chosen at random because it matched the island's name Wight? Whether this is true or not, the location is a genuine one.

Researchers who wish to pinpoint the place more accurately may wish to know that by 1769, there were two windmills operating in the Freshlands area close to the western tip of the Isle of Wight and the Needles rock formation; one on a hill east of Weston Farm and another nearby at Freshwater Green. Neither of these, as enquiries to local historical repositories confirm, seem to have had a farmer called White living nearby. Even so, evidence does show that there was a farm at Headon Hill, also known as Headon Warren, a site that had in earlier times once been a fenced rabbit farm in the charge of a warrener, who lived in Warren House on the Hill. In addition, an Ordnance Survey map for 1793 shows a watermill on the Aftonside of the Freshwater Causeway, together with a millpond.

The mysterious Mr White's name first cropped up in a short article in The (London) Star newspaper less than a week after the incident:-

> *"The officers and seamen of his Majesty's sloop SPITFIRE publicly return thanks to Mr White of South Yarmouth for his attention to the body – This benevolent farmer had made every decent preparation for his internment at their arrival".*

The Salisbury Journal of 3 February 1794, in a tail-piece to the report on the supposed drowning, repeated the story. As we shall see later, it would seem that far from making arrangements for the body's burial, the young person who supposedly discovered the body appears to have wanted little to do with it and ran back to his farm without mentioning the fact to his father.

VEIL OF SECRECY

As stated, though most newspaper reports of the time said that Cook's body was recovered from the Isle of Wight, the authorities at the time appear not to have been too sure. Shortly after the incident, letters were directed to customs officials in the region inquiring whether Cook's body was washed up near Christchurch on the Dorset coastline, a belief that evidently still survives amongst some people today; a point made reference to in modern correspondence with Sheila Caws the Island Heritage Librarian at Cowes who wrote

> *"...I can't find the original enquiry but from memory his body was washed up in Dorset and I seem to remember getting a newspaper report from the Dorset History Centre...*
>
> *I don't think there is any I.W. connection."*
>
> [Isle of Wight]

Another query directed to the archives department of Dorset History Society (Dorset County Archives) asking whether any records were available about Cook's drowning received the following reply in April 2010:

> "*I have, on your behalf, checked our indexes, electronic catalogues, and database of accessioned records, to locate any documents relating to James Cook junior and his death in 1794 but unfortunately, my search has not identified any documents relating to this matter".*

This confirms a note sent from Christchurch at the time of Cook's supposed death, signed by Customs Chief, Robert Richardson who stated in a reply to a question from the Customs authority, that no bodies had been found anywhere in the neighbourhood.

As with most major events or scandals that take place in modern times, reports in press columns immediately following the incident are sketchy. In fact the newspapers of the time were curiously united in failing to give any real explanation of what led up to the sinking of the Spitfire's rowing boat. It's also noticeable that the same newspapers seemed strangely reluctant to publish any follow up reports.

A single piece of surviving documentary evidence shows that Cook (as a commander) had drowned whilst in the service of his country. It was printed barely a year after the incident in *'David Steel's Original and Correct List of the Royal Navy'*.

From that time onward, few printed naval records mention the name of James Cook junior despite his former high rank within the Royal Navy. In fact, following the first reports of his death, his name (as a Royal Navy officer) appears to have been virtually wiped from all history and record books.

As a consequence, the young officer appears to have largely evaded history altogether, so much so that many specialists who have an interest in the maritime activities of his father Captain James Cook senior are unaware that his son, Captain James Cook junior, ever existed. One wonders if the incident of the alleged drowning of the commander of HMS Spitfire may have been deliberately veiled in secrecy, a fact that hasn't entirely gone unnoticed by some vigilant maritime historians. Alan W Smith, for example, a member of the Captain Cook Society, points to the mysterious circumstances surrounding the death of James Cook junior in a series of articles printed in "Cook's Log" the quarterly journal of the Captain Cook Society in 1996.

It's interesting to note that newspaper reports of the time tell us that the young naval commander is said to have taken part in a farewell dinner, leaving the celebration *"immediately after tea"*. If this was an official engagement, then this could account for a body being found wrapped (not *'dressed'* as sometimes reported), in a naval officer's greatcoat; though the Morning Chronicle report of 28 January 1794 did actually state that the body was *dressed* in a full naval officer's uniform. It's also interesting to note that in the same newspaper column where he was described as being *"remarkably serious and melancholy"*, curiously three exclamation marks are then printed after the phrase *"-but his sense of duty prompted him to persevere!!!"* One could be forgiven for thinking that these phrases are merely journalistic, but is there more to it? A question we will examine in due course.

On 30 January 1794, The Times newspaper gave its own version of the story, but revealed little further information, other than that it was a Captain Roberts who informed Cook's mother of the tragedy. This report is also curious for its use of the word 'supposed' in the phrase "*...in which he was* ***supposed*** *to have missed the ship in the night*." Incidentally the town of Yarmouth mentioned in this press cutting is not the well known Great Yarmouth, but Little Yarmouth on the Isle of Wight.

A similar story in the Star newspaper (London) on the same date also uses the word 'supposed' and has the additional information that the mysterious Isle of Wight farmer, Mr White of South Yarmouth had attended to Cook's body. As we have seen, no verifiable account of Mr White's actions appears to have been historically recorded anywhere. Furthermore, local history societies, museums and record offices on the island have no records whatsoever concerning the incident or of Mr White the farmer.

We need to wait for a period of two months to discover possibly leaked information showing that previous press reports may not have told the whole story. On 13 March 1794 the somewhat irritated St James Chronicle or British Evening Post revealed that it had heard from some unknown source that assistance was unexpectedly denied to Cook and his crew during that fateful night and what's more the Admiralty had been sitting on the information for some time.

We may never know if local newspapers were 'fed' or simply agreed to publish an 'official' version of events whilst covering up a greater truth; though it could be argued they were merely attempting to piece together a story from the few facts available at the time. Whatever the truth it would seem that throughout the next fifty years (and beyond) written accounts often stray considerably from the specifics given in earlier newspaper reports.

In 1836 for instance, the book *"An Historical Account Of The Circumnavigation Of The Globe."* published by. Oliver & Boyd, Edinburgh, gives the following version:-

> *"...James who at the age of thirty-one* [he was actually 30 but in his 31st year] *was drowned in 1794, the only son who attained his manhood, displayed much of his father's intrepidity. When pushing off from Poole to join the Spitfire, sloop of war, of which he was the Commander, he was advised to wait until the storm which was raging should abate: - 'It is blowing hard', he replied, 'but my boat is well manned, and has weathered a stronger gale, we shall make the ship very well, and I am anxious to be on board.' He perished in the attempt, along with the whole of his crew".*

This account of Cook's specific spoken words must be suspect at the least, and is possibly a complete fantasy intended purely to add interest to the narrative, because the conversation is not corroborated or repeated in any other surviving record. In short there appears to be no proof whatsoever that Cook ever used those words, though with regards to the weather, a chart printed in the Gentleman's Magazine for February 1794 shows that on the night of the 25 January the weather was indeed classed as stormy. It states that the wind was blowing north easterly with a temperature of 39 degrees.

A Drama Unfolds

Captain Roberts

Roberts (seemingly the same person of that name who informed Cook's mother of the tragedy) was responsible for instigating the inquiry concerning Captain Stephen Watson's negligence in causing the death of Cook's rowing boat crew. Roberts reported a conversation with John Butler, the Mate of The Hound, another cutter also in the service of the Excise service, to the Commissioners of H.M. Customs at Shoreham. Roberts claimed that Butler had told him that the Hound was lying close to the Isle of Wight on the Monday or Tuesday morning following 12 January 1794. At that time crew members belonging to the Grey Hound (a.k.a. Greyhound) Revenue Cutter of Weymouth commanded by Stephen Watson, came on board to pay them a visit. He said the conversation between the men had turned to the recent death of Captain Cook (junior).

After the Grey Hound crew had departed, one of the Hound's junior officers James Paine Bundell informed Butler that the men belonging to the Grey Hound's boat had declared to him and other crew members (of the Hound) that the death of Captain Cook and those with him was the fault of their own captain, Stephen Watson. Charles Storey and William Naylor, two other mariners on the boat also confirmed the same accusations some days later.

It would seem that Captain Cook's crew had rowed or sailed out to the Spitfire, despite the sea being quite choppy. Spotting the Grey Hound cutter which was 'lying too', someone from Cook's boat then hailed them and 'begged' them to give them a rope. It seems likely that the men may have already been in a state of distress because according to the evidence of William Brett, second Mate of the Grey

Hound, one of the men actually shouted *"For God's sake give us a rope."* The wind was described at that time as *'blowing strong from the souther together with some sea running'.*

Stephen Watson

Watson was the surly Captain of the Grey Hound Cutter, in the service of His Majesty's Customs. He was accused by some of his own crew members of wilful negligence in causing the death of James Cook junior and his crew. On 4 April 1794 Watson was officially charged with all the accusations made against him in a document from the Surveyor General's Office. It was signed by John Nugent and A Hammond, who were Surveyors General at that time. The charge put to him read as follows:-

> *To Stephen Watson Commander of the Grey Hound Cutter in the Service of this Revenue.*
>
> *In pursuance of a minute of the Honorable the Commissioners of His Majesty's Customs, dated the 4th of March 1794 in consequence of a letter from Captain Roberts stating an account given by Mr John Butler, Mate of the Hound Cutter in this service, of which the following is an extract, we charge you with the facts recited in the said extract as stated by said Butler. (Viz)*
>
> *He says they were lying under the Isle of Wight when on Monday or Tuesday morning preceding the 12th ultimo, a boat belonging to the Grey Hound Revenue Cutter of Weymouth commanded by Stephen Watson, came on board them with the Mate to pay them a visit, when the conversation turned upon the death of Captain Cook which however this man seemed to wave* [avoid?]. *After staying a short time he left Butler and returned on board his own vessel. Immediately the boat was gone, one of the junior officers informed him that the crew belonging to the Grey Hound's boat had declared to him and the*

people that the death of Captain Cook, and those with him was owing to their Captain, Stephen Watson.

That the boat of the Spitfire as she was going from Poole to Studland Bay where the ship lay, fell in with the Grey Hound cutter then lying too, when they hailed and begged them to give them a rope (the wind blowing strong from the Southward and some sea running), when in rowing up under the cutter's stern, the sweeps [long oars] *which projected some distance, struck the boat and stove in her bow – The man who was at the lead threw them a rope which they caught in the boat, at this instant Captain Watson came upon deck & with most violent imprecations exclaimed 'What are you about', abused the man who threw the rope as well as the others who assisted for their attention to 'such fellows who were only a parcel of damned Men of War rascals'.*

He immediately ordered the foresheet to be let down and stood to sea; the vessel going with great rapidity thro' the water, dragged the forethwart [The front bench seat], *to which the rope was fastened out of the boat; and they heard one of the men in the boat distinctly cry out 'Lord Have Mercy Upon Us' and others bewailing their fate.*

That Captain Watson never attempted to put the vessel about or used the least means to save the unfortunate sufferers.

That on the Spitfire coming up to Spithead, Watson immediately went to sea with his cutter, a proof in Butler's opinion of a guilty conscience.

To which matters you are hereby required to make a plain & distinct answer in writing and annexing this charge thereunto.

Surveyor Generals Office 4th March 1794 [signed by Nugent & Hammond]

Verbal evidence subsequently given at the enquiry repeated much of what was in the official charge, revealing that when the Spitfire's boat came up under the cutter's stern, the sweeps [long oars] which projected some distance outwards had indeed struck the boat and subsequently stoved in her bow. The man who was at the lead, [a lead weight attached to a line used in sounding depth or measuring speed] then threw them a rope which they caught hold of. Captain Watson, it was confirmed, came on deck in a livid mood and angrily exclaimed "*What are you about!*"

He was said to have abused the man who threw the rope as well as the others who had assisted him, for paying attention to what he considered *"Damned Men of Wars Rascals"* who he said could "*Go to Hell"*. He had then allowed the ship to set off rapidly through the waves, apparently tearing the front out of Cook's open boat as described in the charge. Some sailors gave evidence that they had distinctly heard one of the men in the boat cry out "*Lord have mercy upon us"* while the others verbalised their distress.

As stated in the charge, evidence was given by some that despite the danger to Cook and his crew, Watson made no attempt whatsoever to turn his cutter around, or to save the unfortunate sailors in the Spitfire's boat. It was also claimed that when the Spitfire, the rowing boat's mothership, came up to Spithead, Watson immediately turned his cutter around and headed for the open sea as Butler had described. Throughout the enquiry, Captain Watson continued to insist that there was a conspiracy against him. Whether or not this was the case, it would seem that regardless of any other facts, he was intensely disliked by many of his crew.

John Butler

Butler was Mate of the cutter Hound (not to be confused with the Grey Hound). He gave his version of the events signed by his own hand:

> *"On the 26th January being at anchor at Spithead near the Greyhound in the evening, Mr Carter the Chief Mate*

of the Greyhound with some of the crew came on board the Hound and after their leaving of them James Paine Bundell one of the departed mariners, as did since, Charles and William Naylor two of the mariners, related to me what the Greyhound people had been saying respecting the death of Captain Cook which was as follows, that they was cruising with the Greyhound within Poole Bar and blowing very strong. The Spitfire's boat came to the cutter and ask'd them to give them a tow to their ship in the bay, at the same time one of the sweeps that was lying over the cutter's stern by her pitching, fell as they supposed on the boat and broke, in which the man that was heaving the lead, threw it and threw a rope to the boat which they made fast in the boat. Captain Watson at the same time came on deck and looking over the topsail told them to get forward (by saying it was nothing but a Man of War boat and they might go to hell) and then shouted go to the foresail, by that the cutter had so much way turned away, thought as they suppose'd the rope was fast to on which their cry'd Lord have mercy on them and they never saw them afterwards."

This evidence was referred to a number of times during the Surveyors General enquiry as '*the paper marked A*'

HABEAS CORPUS – Where was the body found?

According to the evidence of people who were on the shore on that early morning at daylight, the wreck of Cook's boat was discovered keel upwards and *"on a ledge of rocks called the Warden"*. They stated that 300 or 400 yards to the west of the boat, a body was discovered sometime afterwards. It lay between two rocks, wearing or wrapped in a gold laced tunic. Nearby were various items from the boat, including a number of hats, but there were no other bodies. A note appended to the report stated that the knees of the body (described as Captain Cook's) were *"very much rubbed which must have been either against the boat or the rocks"*.

The question of whether Cook really drowned, or even that his (or any) body was found on the rocks at Warden Point or elsewhere, must be called into dispute when we look at the response given by various officers of the Customs Service in reply to questions put to them concerning the supposed drowning.

Poole Custom House

On 8 March 1794 Poole Custom House communicated:-

> *"Having in obedience to your honours directions signified by the Secretary's letter of the 7th instant, No 25, made every inquiry in our power whether any person was saved belonging to the boat in which the late Captain Cook was unfortunately drowned. – We humbly beg leave to report that we cannot learn of any one and that we have reason to believe that every person who went from hence in her perished"*

On 14 March, John Lander (Collector of Customs) and Edward Allen (Comptroller) at Poole Custom House wrote a letter to Burrow & Hammond, Surveyors General. Their letter revealed some interesting information, namely that Edward Allen's own brother was one of the men on the stricken boat, the only evidence anywhere it seems that reveals the surname of any of the lost mariners who accompanied Cook that night:-

> *"Gentlemen, in answer to your letter of the 13th instance, we beg leave to inform you that the Comptroller (whose brother was one of the unfortunate sufferers with Captain Cook) declares that he was present when Captain Cook*

and his boat's crew went off from Poole, and that he is positively certain that all of them were perfectly sober at that time, and that the supposition that some of the crew were inebriated and obliged to be tumbled into the boat is totally groundless. We are respectfully Gentlemen, your most dedicated and humble servants." [Signed by Sander & Allen]

Another letter signed by L Lander the Collector at the Custom House in Poole on 17 March 1794, was addressed to Edw. Burrow & A. Hammond at the Surveyor General's Office London. It was obviously in reply to a second inquiry asking if Cook's boat had come ashore at Brownsea Island (which constitutes the largest island within Poole Harbour). In this letter, the term 'Sitter' refers to an Officer in charge of Customs boats. Unfortunately the letter asking the question is not contained within the documents held at the NA.

"Gentlemen, In answer to your letter of the 15th instant, I respectfully beg leave to inform you having made the enquiry therein directed, I find on the examination of Mr Robert Arrowsmith the sitter at Brownsea, Captain Cook's boat did not come on shore there, but in the evening of Captain Cook's departure from Poole, a boat which was supposed to belong to the Spitfire passed Brownsea Key and stood out of the harbour about 9 o'clock."

Admiralty Office

In a letter from the Admiralty Office dated 11 March 1794 (signed by what appears to be P Stephens or Stephenson), it was stated that:

"Having communicated to my Lord Commissioners of the Admiralty, your letter of the 7th inst. inquiring the request of the Commissioners of the Customs, to know whether any person was saved belonging to the boat in which the

> *Late Captain Cook was unfortunately drowned and if so, the name and residence of such person or persons, I am commanded to acquaint you, for the information of the board of Customs that by the accounts received at this office, it does not appear that any of the said persons were saved."*

Portsmouth Custom House

The finding of no bodies whatsoever appears again in a letter from the Custom House in Portsmouth addressed to the Surveyor General at the Customs House in London and dated 27 March 1794:

> *"Gentlemen, in answer to your letter of yesterday's date we acquaint you that we have made enquiry respecting the men who were in the boat at the time Captain Cook of the Spitfire was unfortunately lost, but cannot learn that any of them has been found."*

Cowes Custom House

A dispatch from Cowes Custom House for the same date, 27 March, replies in the same vein except that it states that no bodies were found *"except that of Captain Cook himself"*, repeating an earlier letter from 11 March that stated the same facts together with the point that two hats supposed to belong to the crew had also been found.

On 31 March they followed up with a longer letter indicating that they had visited the place where the body was found in order to make enquiries of local people who were on the shore on the morning the body was found. Their findings revealed that the boat was discovered around daylight, keel upwards on a ledge of rocks called The Warden, the mast 'unstepped' and hanging alongside by the halyards [the ropes used to haul the sail]. About 300 or 400 yards to the westward of the place where the boat was discovered, a body was found some time afterwards, lying face downwards between two rocks high up on the shore, and near it a golden laced hat at a distance of 100 yards to the westward of the body. The ceased-up sail was also picked up. An oar, boat hook and some other sundry articles were collected on the shore both before and after the boat was first discovered, including four or five hats. Two or three more hats were later gathered up at different parts of the shore *'...at no very great distance from the boat'*, though one of the hats shown to them was evidently not considered to have belonged to a sailor.

They interviewed both the 'sheep farmer' who found the body and another un-named person:

> "*...We examined another person on whom some suspicion had fallen, that he had before seen and robbed the body as he had been seen to pass very near the place where it was found very early in the morning – Whether from conscious guilt or from fear and ignorant timidity at being again examined having previously been suspected and examined upon it by a neighbouring magistrate who could find no ground for committing him, he was very unwilling to give any direct answer and would not for a long time say whether he had or had not seen the body, however he did at last acknowledge he had seen the legs of a man on the shore early in the morning which so frightened him that he immediately ran home and did not go near the place again the whole day, and what is a little singular and we think gives great cause for suspecting him of acting improperly in the business is he says he did not mention to anyone who he had seen, not even his father with whom he afterwards went to plow – We find that there was no money or watches in Captain Cook's pocket when found by the farmer, except a guinea in a letter for one of his sailors...*"

The statement goes on to repeat that no other bodies were found and that the strong ebb tide must have taken them out to sea, making the suggestion that the Captain possibly clung to the upturned boat until it was washed ashore where he probably died from fatigue. The fact that a letter containing a guinea, a coin worth one pound and one shilling, which was quite a sum at that time, that was intended for one of the sailors would seem to indicate that the body had not been robbed at all. Could it be that the body on the beach wrapped in the captain's coat wasn't actually Captain Cook and may perhaps have been the sailor whose name was on the letter referred to? What is certain from a coroner's letter is that the man had not actually drowned, but had been alive when he came ashore, a fact noted by the Cowes Custom House staff.

> "*...The body when found was in a length way position, the head towards the land, whereas had it come on there a corpse, it would probably have been found broad side to the sea. – The wind which at 4 o'clock in the morning was S W blowing, though shifted about, as we were informed by a fisherman to W N W whereas which was likely to drive a boat coming in at the Needles, upon the Warden Rocks. In addition to the foregoing particulars we beg leave to enclose a letter the Collector has received from the Coroner who took an inquest upon the body and who has requested to commit to writing such particulars as he could recollect of the evidence produced to him. With respect we remain your most obedient servant* [Signatures which seem to be] Bedlington, Arnold and Ward."

Robert Clarke (Coroner)

It would seem strange that an inquest was carried out on the death of Captain Cook junior, but apparently not on the deaths of the other lost sailors. There is nothing to identify exactly where or when the inquest took place so it's not possible to know how it was conducted (e.g. was the body examined and identified, or was the matter simply heard by a jury with the aid of witnesses?). The letter received from the

Coroner mentioned in Cowes Custom House sent around 31 March has survived. It was signed by Robert Clarke but was not dated, and it isn't clear who it was addressed to (i.e. the Customs Officers or the Enquiry). In it we discover that the man identified as Cook had not actually drowned, but had probably died of fatigue after reaching shore. We also learn that the man wasn't wearing a Captain's great coat, but simply had it wrapped around him. The letter is transcribed here in full:-

> *"Dear Sir, I have been recollecting the evidence given upon the inquest of Captain Cook, and the only circumstances which appeared to the jury, were that as a man was tending his flock upon the high sands at Freshwater, he saw something floating in the water near the shore, this induced him to go down, and in coming to the water he found it to be a basket filled with vegetables which he dragged on shore under the expectation of another. He then walked on and discovered a hat and hard by, too* [sic] *rocks out of the reach of the waves he saw the unfortunate man dead, but not stiff. The posture in which he was found was singular. He was reclining on his arm in a position of sleep with a Great Coat wrapped completely round him. The boat was found at about a quarter of a mile distant. I think but am not positive, with its keel uppermost. Indeed I am very inclined to think that the boat upset and that Captain Cook contrived to get upon her and was driven on shore but exhausted by his efforts and from the intense cold that his power of exertion failed and that he laid himself down in the position he was found. There was a violent contusion over one eye but it* [sic] *impossible to say how he received it. – There was a report circulated at the time, and I believe with some foundation that a man passed by the body on that morning at day break and found the boat, but he denied that he saw the body. – It was nine o'clock before the body was*

> *found, at least known to be so, and then it was not stiff, so that if it was discovered at daybreak and proper means had been tried, it is sure than probable that he might have been saved. Mr Rushworth to whom I represented this matter took every hand to get into the truth of this, but could make nothing of it. I am* [signature] *Robert Clark."*

The above re-confirms that the man found on the shoreline was alive when he reached the shore, though the body seems not to have been identified in the normal fashion, i.e. by facial recognition, jewellery on the body or perhaps the location of a scar. Instead it seems the officer's greatcoat wrapped around it was sufficient to identify the body at the scene of the incident, disregarding the fact that it may not have been Cook but another of the crew. Could the body have actually been wrapped in the greatcoat by Cook himself before he departed the scene, leaving the other man on the shore? Considering the North Yorkshire legend about Cook's survival could it also be that Cook seized the opportunity to make his escape and to desert the Navy forever as the stories would have us believe?

Southampton Customs House

Southampton officers also appear to have had a similar letter asking if they knew of any bodies being found. They wrote:

> *"Gentlemen. We have never heard the least of the unfortunate Captain Cook or his boats crew since the accident happened, but in case we should, you shall be immediately acquainted with it as directed in your letter of the 26th instant. The letter was signed by William Smith and what appears to be S Morton."*

Lymington Customs House,

In a similar vein Lymington officers advised oddly that *"...**neither*** [My emphasis] *of the men who were in the boat at the time* [when]

Captain Cook of the Spitfire was unfortunately lost have been found in this part of the coast." Does this indicate that they assumed only TWO sailors were in the rowing boat when it was involved in the mishap and that nobody was found?

Christchurch Custom House

A communication from Christchurch signed by Customs Chief Robert Richardson said that no bodies had been found there or anywhere in the neighbourhood.

Richard Routh (Cook's friend)

It appears that those conducting the enquiry left no stone unturned. The set of documents in the National Archive collection contain a letter from Richard Routh who described himself as a friend of James Cook junior. This would seem to be the same Richard Routh who, like Cook, was born in Yorkshire on 11 December 1749, but went on to make his mark in life as Collector of Customs in Newfoundland. Because of the rebellion in the American colonies, Routh returned to Poole, England in 1778, and made his home there. He too met his death in mysterious circumstances whilst a passenger on a voyage from England to Newfoundland on the 16 gun H.M. Sloop of War The Fly which vanished without trace, never to be seen again.

Routh, like Cook, was evidently a well-liked and generous man who was known for giving parties for his friends. His letter solves the mystery of where Cook was dining prior to boarding the open boat which was to take him back to the Spitfire

In Routh's letter dated 1 April 1794 he confirmed that Cook had been asked not to get in the boat that evening because of the worsening weather conditions: -

> "....*Captain Cooke was a friend of mine & with Capt. Yeo the Regulating Officer of the Impress dined with me the day preceding his death. We endeavoured by every*

argument in our power to dissuade him from returning that night to his ship the Spitfire, which laid in Studland Bay as it appeared to be dark and tempestuous, but no argument could divert him from his purpose – I had my fears about him, particularly towards bedtime & was anxious to hear something of him but nothing transpired till the next evening, when a boat came from the Spitfire & I was informed he was missing – The following day I received a letter from the Lieutenant of the Spitfire acquainting me that he learnt by an express that the body of a Master & Commander had been found on Warners Ledge; this left no doubt of his unhappy fate, which agitates me more than I can express, a day or two afterwards I was told that Captain Cook was found between two rocks, that the boat and oars were near him & a basket with some oranges also on the beach – that he appeared to be scarcely dead when first discovered & that he had been robbed of his watch & money – Mr Tyander, the Master of a Sloop coming from Portsmouth went on shore & saw Captain Cooke – I talked with him yesterday, he says he was dressed in his uniform & had a flushing coat [A loose-fitting, hooded garment similar to a modern 'hoodie'] *over it. That he did not appear to be drowned, in short that he could not have been drowned, having no water in him; that his forehead was disfigured a little & the left side of his face rather discoloured – that he supposed he was weak and much fatigued & in getting from the boat might have fallen & hurt himself – He confirms the circumstances of the boat being found near Capt. Cook, but did not recollect if the oars were in it – I saw also yesterday the Rev. Mr. Cooth who tells me he was visiting on the Isle of Wight & saw the person who first discovered Capt. Cook who said he was much alarmed & went to a short distance for some assistance to go with him to the spot. Mr Cooth relates that he understood the boat was found some time that morning before Capt. Cook was discovered, that the body was taken to a Farmer White's house, as the person who discovered him was fearful of distressing his wife, their son having been lost in this way sometime before. ---- Whether any of the circumstances are material I don't know, if they are I believe they may be relied upon, or if any doubts should arise you may be able to make enquiry from the people who live near the spot. - Warner's Ledge is I believe opposite Hurst Castle.- You may be perfectly assured that Capt. Cook was sober when he left my house that evening – there was no wine drank that could possibly make him otherwise. I have every reason to believe that he embarked immediately after he left me; He was a man of great resolution, steady to his purpose & it appears to me did not look upon the situation he was about to be in as hazardous, though I repeatedly wished to impress him with my opinion that it was."*

William Turner (Grocer & Cook's Friend)

Turner, another friend of Cook, penned a letter on 1 April 1794 after Captain Steven Watson, the man accused of being responsible for the death of the boat crew, asked him to repeat a conversation that took place in Turner's grocery shop situated *"...on the point, Portsmouth"*. Portsmouth Point became known as "Spice Island" because of its seedy reputation for prostitution and was the place where sailors were regularly impressed for Royal Navy service by the Press Gang. Watson asked Turner what he recollected about the conversation which had taken place in his shop concerning the death of Cook, in the presence of Admiral Boyer [Bowyer], Captain Bazely [Possibly the later Admiral John Bazely 1740 – 1809] and other officers. Turner wrote on 1 April 1794:

> *"I write this in answer to your request asking me to inform you what I recollect of the conversation that passed in my shop in relation to the melancholy fate of my worthy friend Capt. Cook which as near as I remember is as follows. Having that morning been told that Admiral McBride had sent an account from Cowes of the loss of my friend, I was of course very much affected and made all the enquiries in my power in hope of proving it false, and as you came into my shop at the time I mentioned it to you, in answer in which you said you feared it was too true, for in the night the incident was said to happen a boat with six men and three sitters came alongside your cutter, and by the description I gave you, supposed it to be Captain Cook, standing up, who hailed you, on which you gave them a rope which they held to, whilst they either set or altered their sail but whilst you were attending to the necessary care of your vessel you supposed they had accomplished what they wished & let go the rope and proceeded where they were going for you neither saw them quit the cutter or anything of them afterwards, but as a most violent storm came on shortly afterwards you supposed it very improbable they could outlive it. I recollect Admiral Bowyer was with me at the time of your conversation, by an answer he made that if officers thought proper to drown themselves they had a right to it but it was sorry for them to risk their men's lives."*

In a postscript he appended a sentence to the letter saying: - *"I realise I am as correct as possible in the aforesaid recollection of what passed as my Clerk who was present at the time remembers the same."*

It's interesting that the accused Captain Watson of the Grey Hound claims to have recognised Captain Cook from William Turner's description, particularly as, according to his crew's evidence, Watson was below decks when the boat was first encountered. Furthermore he didn't come up from below decks until after the rope had been thrown. It is also remarkable that no mention is ever made of him personally being asked for a rope to be thrown during his official examination at the Custom's Board inquiry.

Though only coincidental, the fact can't be dismissed that if Cook was reported as being in a *"remarkably serious and melancholy"* mood on the evening of his final departure, and Admiral Bowyer had made mention of officers drowning themselves, had Cook said something to the gathering prior to his departure from Routh's house that suggested that he felt like killing himself for some reason? The letter which he evidently wrote to a certain lady, mentioned in newspaper reports at the time, may have revealed all, but it appears to never have surfaced again. In reality Admiral Bowyer may of course have just been referring to Cook's stupidity in setting off for his ship in an open boat in the terrible weather conditions that evening, but the truth of the matter must ever remain a mystery.

LOST TO HISTORY? – The mysterious lack of records

We know that the evidence of the day was accepted as truth by the inquiry, namely that the remains of Cook's rowing boat was eventually washed ashore 'within Warden Lodge' and that nearby lay a body which was claimed to be that of Captain Cook, though according to later evidence he showed no signs of having drowned. Some of the boat's oars and a number of hats were discovered but no other bodies.

The location 'within Warden Lodge' appears to be near Warden Point which was at that time devoid of any buildings except one. The scene was still the same a year later in 1795 when the painter J.M.W. Turner sketched it and described the area as one which, apart from the views across the Solent, had only a lonely cottage and nothing else nearby.

It seems very strange that no one seemed interested in the fact that the bodies of none of the sailors aboard the rowing boat (other than one supposedly identified by his clothing as Cook) had ever been found and that no search was made for them. This is indeed a very odd situation because at least one letter from a Customs Office appears to claim that *none* of the bodies of those in the boat were ever found. More to the point, no list of the names of missing sailors appears to ever have been compiled or published; and as has been stated, no record of the incident was made public apart from the original 'press release' stories issued within the first few days after the event.

The Naval Chronicle of 1799 whose mission statement' declared:-

> *"We shall endeavour to make the Naval Chronicle a useful and interesting library of itself to seamen, and an acceptable work to everyone who partakes of the glory acquired by our brave countrymen ... Our leading principle will be to adhere strictly unto truth".*

Oddly its contents made no mention whatsoever of the incident and didn't mention Captain Cook in its list of drowned or deceased officers, nor was the incident made reference to in any of the naval memoirs contained within it.

Because civil records were not introduced until 1837, no death certificate exists for James Cook but surely it would be assumed that the demise of a prominent naval commander, himself the son of a famous hero of his day, would have prompted widespread discussion both in newspaper columns and naval circles everywhere. As such it should follow that his life and death details would be recorded extensively in naval records. Extraordinarily this appears not to be the case. An extensive search reveals that local, regional and national archive record offices appear to be devoid of any such information. More inexplicably, the log of the Spitfire itself has no mention whatsoever of the drowning of its own commander and crewmen, (even though they were supposedly rowing their way towards it to take up their duties on the night in question!)

Katherine Weston of the National Maritime Museum archives department confirmed this in an email in June 2015:-

> *"I have searched through the log book for the Spitfire (including the months either side). Unfortunately, there is no mention of the loss of Cook. It does however, record that the boat was moored close to Yarmouth and the Isle of Wight around this time and also mentions that the weather was very stormy with gale force winds toward the end of January".*

Oddly, in the absence of any other documentation, it seems that an officer's coat was the sole means of identifying Cook's body. One might be inclined to believe that it should have been members of the crew of Cook's own ship, the Spitfire, who were asked to identify him when the body was collected to be delivered for burial. However, by that time the corpse may already have been wrapped in a shroud or placed in a coffin. And as the Spitfire's ship's log doesn't record the drowning of its Captain or his crewmen, the picking up of the body, or even his death, what are we to believe?

If this situation is not enigmatic enough, perhaps more intriguing is the apparent significance of the letter that Cook wrote to the mysterious unknown lady on the night he died. We might also ask why the contents of the letter were not even discussed at the enquiry, despite its perceived importance. The Morning Chronicle first mentioned it on 28 January 1794 and the story-line was repeated word for word the next day in the Public Advertiser or Political and Literal Diary (London) with the additional intriguing paragraph:-

> *"We are sorry to learn, that the report of the death of this young Gentleman, who was Commander of the Spitfire sloop of war, and only surviving son of the celebrated Captain Cook, is confirmed by* ***an official letter to His Majesty****".*

An official letter to the King would surely seem a certain way of verifying once and for all that James Cook junior actually died that night but unfortunately like all good mysteries there is a catch. Regrettably there is no trace of this supposed letter to be found anywhere in the Royal papers, a fact that was verified in June 2010 by the Registrar.

Registrar of the Royal Archives:

> *"Thank you for your letter of 1 June enquiring about a letter sent to King George III reporting the death of Lieutenant James Cook in January 1794. I have read through all of the King's surviving official correspondence in the Royal Archives for that month, but I regret this does not include the letter in question..."*

Of course it is possible that the letter has been lost over the years, but this doesn't account for the total lack of information about the incident in prominent archive collections elsewhere.

National Museum of the Royal Navy:
A inquiry directed to the Royal Navy Museum in May 2010 resulted in this response from Library assistant Mrs H. Johnson,:

> *"James Cook Junior passed for Lieutenant in May 1782 but I can trace no other appointment for him apart to HMS Spitfire. The weather was particularly bad that January but there is nothing in our sources that relates to his loss."*

Isle of Wight County Record Office: Considering that Cook's body was supposed to have been washed up on the Isle of Wight, one would expect that their County record office would have some archived account of the incident. However this is not the case. In fact Richard Smout, Heritage Service Manager for the Isle of Wight County Record Office replied in April 2010 that

> *"The churchwardens' accounts for 1793 and 1794 for Freshwater, one of the more likely places for James' body to have been washed ashore, contain no reference to this incident".*

Further enquiries to relevant repositories elsewhere drew a similar blank. Neither the Naval Records Society, nor Portsmouth City Archives could shed any light on the subject and it would it seem that apart from references to his drowning, his career details also appear to have been expunged from Royal Navy lists of officers, biographies and similar books where such a prominent Naval Commander would expect to be found.

All in all, a very mysterious situation!

Recollections & Mercurial Truths

Before we examine the official enquiry into the incident, it's interesting to look briefly at how various accounts of Cook's death have appeared in print. Below is a chronological selection of them. They show the quite varied ways in which Cook's supposed death was actually reported. As will be seen, some vary in only small details whilst others disagree totally with the information given by other sources to the extent of repeating earlier misconceptions. Some contain total inaccuracies.

JAMES SADE – (An old neighbour of the Cook family at Mile End). c1794

James *"...was lost agoing on Board his ship which was under sailing orders and the evening being very dark and windy he and six people was overset and lost."*

GENTLEMAN'S MAGAZINE, Vol. 64, Page 182 – Obituaries of Notable Persons –February 1794

"... Aged 30, Capt. Cook of His Majesty's ship Spitfire, drowned in Poole-harbour. He was the eldest and only surviving son of the celebrated navigator. The boat is supposed to have missed the ship in the night, and afterwards upset on one of the ledges of rocks near the Isle of Wight, as the body was found near a mill. The boat had drifted close to the place where the body was picked up; but no part

of the boat's crew, which consisted of the coxswain and seven men, have been found. The captain, had he lived, bid fair to have been an ornament to his profession. His body was brought to Cambridge, Feb 4, and buried in St Andrew's church, in the same grave with his brother, who died there of a fever, in his 18th year, Dec 28, 1793. One brother was drowned in the Thunderer, **,** *in a storm; and two daughters were married to naval officers, who were both drowned."*

It's interesting that this isn't the only reference to James Cook junior having married sisters as will be discussed later.

THE READING MERCURY, 3 February 1794

"Captain Cook, of the Spitfire fire-ship, so unfortunately lost with his barge crew off Poole bar, was the eldest son of the celebrated navigator, Captain Cook, killed by the natives of one of the Sandwich Islands."

BURIAL REGISTER OF ST. ANDREW THE GREAT, CAMBRIDGE, 5 February 1794

"James Cook son of James Cook the Circumnavigator & Elizabeth his wife who lost his life as did the whole boat's crew in the passage to his vessel the Spitfire which lay off the Isle of Wight was buried in the same grave with his brother Hugh within this church."

EXTRACT FROM CUSTOMS COLLECTOR TO BOARD LETTERS BOOK 1792 – 1794. 11 March 1794

"We have made the enquiry by your Order of the 7th Inst. and cannot hear that any Person was saved belonging to the Boat in which the late Captain Cook was unfortunately drowned or any of the Bodies except the Captains taken up on the Island Shore. Two Hats supposed

to have belonged to some of the Boats Crew were picked up much about the time the Body was found."

AN HISTORICAL ACCOUNT OF THE CIRCUMNAVIGATION OF THE GLOBE, – AND OF THE PROGRESS OF DISCOVERY IN THE PACIFIC OCEAN, FROM THE AGE OF MAGELLAN TO THE DEATH OF COOK." – Harper & Brothers, New York, 1837

"... James, who, at the age of thirty-one, was drowned in 1794, the only son who attained to manhood, displayed much of his father's intrepidity. When pushing off from Poole to join the Spitfire sloop of war, of which he was the commander, he was advised to wait till the storm which was raging should abate: — "It is blowing hard,' he replied, but my boat is well manned, and has weathered a stronger gale; we shall make the ship very well, and I am anxious to be on board." He perished in the attempt; along with the whole of his crew... The widow... Such was her sensibility, that on receiving tidings the death of her son James, in the vain hope of banishing from her mind the recollection of her losses, she committed to the flames almost all the letters she had received from his father."

LE KEUX'S MEMORIALS OF CAMBRIDGE – VOL. II, 1842

"...James a commander in the navy who lost his life on board the Spitfire sloop of war, in going from Poole to Spithead in 1794."

AUSTRALIAN TOWN AND COUNTRY JOURNAL, 31 May 1890

"... in the autumn of 1793 promoted to the rank of commander, was, while with his ship at Poole, in Dorsetshire, appointed to the command of the Spitfire, Sloop-of-war. On January 24, 1794, he received from

Captain Yeo, commanding officer of the station, his letters and orders to take command without delay. He started immediately in an open boat, manned by sailors returning from leave, to sail from Poole to Portsmouth. It was in the afternoon. His boat was rather crowded. There was a strong ebb tide, and a fresh wind; it was growing dark. This was the last seen of James Cook, the younger, for he never reached his ship. What happened will never now be known. His body, with a wound on the head, and stripped of all his money and valuables, was found on the beach at the back of the Isle of Wight. The boat was also found broken up; but no trace of any of the crew was discovered. Perhaps they were drowned. Perhaps they murdered the captain, made for the island, laid his body on the beach, broke up the boat, and dispersed."

NOTES & QUERIES, October 1892

"....Mrs Cook never ascertained whether he lost his life by the treachery of the boat's crew or whether he was drowned in the ordinary course of the passage from Poole..."

ARTHUR KITSON (From his book The Life of Captain James Cook), May 1907

"... James, rose to the rank of Commander, and in January 1794 was appointed to H.M. sloop Spitfire. He was at Poole when he received his orders to join his ship at Portsmouth without delay. Finding an open boat with sailors returning from leave about to start, he joined them. It was blowing rather hard, and nothing was ever heard of the passengers or crew, except that the broken boat and the dead body of the unfortunate young officer, stripped of all money and valuables, with a wound in the head, was found ashore on the Isle of Wight."

Myths, Mysteries & Misconceptions

By the age of thirty, James Cook junior had ample time to acquire a number of female friends and in sailor's fashion no doubt had many opportunities to sow his wild oats whenever he went ashore. If this is so, it must be at least likely that the genes that James Cook junior inherited from his father could still be found at various locations. If so, then contrary to accepted wisdom, there may well be direct descendants of Captain James Cook (senior) living in the world today. Regardless of this fact, we shall examine the compelling rumours that James junior had a legitimate child or children by way of a marriage in the Lilling area of North Yorkshire. If true, then James Cook lived for many years after his supposed death on that stormy night in January 1794.

Some modern writers have made claims that traditional accounts of James Cook RN and his family are full of errors and that the works of even his most acclaimed biographers are based on handed down information rather than dependable research. In the case of someone as famous as James Cook RN, few are willing to challenge what has now become written history, though nods and winks have been given in various books and papers that all is not as it seems regarding James Cook senior. Because these affect the genealogy of James Cook junior, we should at least address some of these accusations, if only to allow other researchers to investigate and correct inaccuracies at some time in the future.

THE FUNERAL MYSTERY

If the burial of Captain James Cook junior took place as reported in newspapers of the time, it was certainly a quiet affair. There appears to be no mention of the actual event anywhere, no lists of prominent colleagues or family mourners and no mentions in local history guides, other than transcripts of the epitaph and repetitions of the story of his burial. With the absence of any written accounts, the first step in confirming that the burial took place was to locate the burial registers in Cambridge.

In reply to an enquiry to Cambridgeshire Archives, the Archives Assistant Yaye Tang kindly forwarded a transcription of the burial in St Andrew's burial registers

> *"'Feb 5 1794 James Cook [BT Captain 30] son of James Cook the Circumnavigator & Elizabeth his wife who lost his life as did the whole boat's crew in the passage to his vessel the Spitfire which lay off the Isle of Wight was buried in the same grave with his brother Hugh within this church.' - The above burial register lays aside any doubts that James was buried anywhere other than at St. Andrew the Great… Please note that civil records were not introduced till 1837, therefore no death certificate exists for James Cook."*

Whilst the burial is officially confirmed, it would seem a particularly imaginative entry, not only fully describing how James Cook died but also giving details of his ship and his brother Hugh who was already buried there. Whilst this could be normal for St Andrews church, the description certainly goes far beyond the normal brief burial register entries found in most churches, almost as if it was intended to confirm the details written in newspapers at the time, should there be any doubters.

Given the Yorkshire legend of Cook's desertion and survival, we are left with three questions.

(1) Was any body at all contained within the coffin (i.e. could it have been a part of an elaborate cover up, and if so was Elizabeth Cook his mother knowingly or unknowingly a part of the conspiracy?)

(2) Could someone other than Cook be buried there? I.e. the person found on the beach who was misidentified as Cook.
(3) Was James Cook actually in that coffin?

We do know that Mrs Cook was naturally bereft at the news that her eldest son was gone from her life, but whether it was because she knew or thought that he had died, or because she had heard he had disgraced her by deserting the navy and herself forever, may never be known.

A document held by the National Library of Australia, Canberra, (Nk 9528), tells us that:-

> *"E. Honeychurch wrote to Mrs McAllister, 12 September 1794, about her, that the news of the drowning 'quite overcame her, and she has not been able to come down stairs, or eat a bit of bread since, within these few weeks she has eat a small bit of Veal, or Lamb, or a little Fish, since which she has thought herself rather better tho' she has two fits every day Night and Morning and they hold her an hour, and I am afraid they will never leave her, it is a long time to be in such a state...."*

It is clear from an obituary in the Cambridge Chronicle (and other publications) that Elizabeth Cook was later buried in a vault (i.e. not a grave) with her children, so in theory it should be possible to discover using DNA technology if the body of the man that was claimed to be James Cook and who lies with her, is actually her son. Unfortunately the exact location of the vault is not positively known, other than it is described as being '...in the middle aisle of this church'. Intriguingly, Whitby, Yorkshire, Museum holds a brooch that belonged to Elizabeth Cook. It contains a lock of unidentified hair (her husbands?), which in theory could be DNA tested to see if it matches that of anyone who claims direct descent from Cook

In April, June and September 1992, the Cambridge Archaeological Unit from the University of Cambridge carried out excavation work within the Church of St Andrews the Great. Though three brick burial vaults were discovered across the central aisle, they were not identified and no other work was undertaken to search for other vaults. The official report into the archaeological investigations states that:-

> *"...As was expected, inhumations were encountered. These were also recorded archaeologically but were only investigated when they lay within the area of investigation. The excavation was not extended to recover the whole skeleton, as it was envisaged that this would only lead to the exposure of more skeletons which would also then need to be followed. The excavated bones were not collected for further study but retained for appropriate burial by the church authorities..."*

Fuller details are contained within the units report but as would be expected they do not concentrate on the identity of those buried there. An email from the unit's office confirmed that the Director *"... informed me that they didn't find a grave for Elizabeth Cook and locating the grave was not one of the objectives of the work the CAU did at St. Andrews the Great."*

A QUESTION OF LEGITIMACY

It has been hinted by at least one author that Captain James Cook (senior) may have been born illegitimately. To examine this possibility we should perhaps begin with James Cook senior's early days when he was working with his own father on the estate of Thomas Skottowe, a gentleman farmer. It was Skottowe who paid for James Cook to attend the fee paying Postgate school (a privilege that evidently wasn't extended to the other Cook siblings), and it was

Skottowe who provided cash for the young scholar's textbooks and other requisites. Why his father's boss should be so magnanimous is unclear. Perhaps he was just a genuine caring employer who saw some intellectual spark that should be encouraged in his employee's son, though it has been speculated in the book 'Captain Cook: Master of the Seas' by Frank McLynn, that James Cook RN could have been the illegitimate son of Skottowe, supposedly without the knowledge of the famous mariner's declared father.

A MATTER OF CLASS

According to most accounts, James Cook senior's wife Elizabeth Batts came from a humble background. Her parents, it is said, ran the dockside Bell alehouse at Execution Dock, Wapping and though it is not precisely documented, it's implied that Cook senior probably met his future wife whilst drinking within its walls.

Other accounts insist that Elizabeth was from a wealthy well-connected background. A third account given by a member of the Cook family states, as we shall see, that she was a clergyman's daughter. If some of these reports of her background are true, then it's likely that Elizabeth may not have even lived in the area at all, only lodged there to comply with church regulations prior to reading the banns before her marriage.

According to the London Borough of Barking and Dagenham Local Studies Information Sheet No.22, Elizabeth Batts was thirteen years the junior of her husband, *"Elizabeth Batts has no known connections with Barking, so historians have long been puzzled as to why she and Cook married there."*

In 1994 Herbert Hope Lockwood, in an article 'Captain Cook and a voyage of discovery around Ilford' in the Barking and District Historical Society magazine, presented evidence suggesting that at the time of her marriage, Elizabeth had been staying with family friends James and Ann Sheppard in their house at Croucher's Yard on Ilford High Road. Ilford was at that time within the parish of Barking. Supporting this theory is the fact that the officiating vicar at the marriage, George Downing, served as Chaplain to Ilford Hospital Chapel.

Whatever the truth, it is generally accepted that Elizabeth's mother was Mary, a daughter of Charles Smith, a leather currier from Bermondsey. Some sources say Mary was a daughter of James or John Batts, son of Richard Batts a seaman from Bermondsey and his second wife Mary Smith. Unfortunately for Mary and Elizabeth, their beloved husband and father died in 1742. No doubt in need of a breadwinner, Mary married again to John Blackburn at Shadwell in July 1745. It is uncertain whether John Blackburn, like Cook, also had Whitby connections. The surname is certainly a local one in Whitby records where the family of Blackburn can be traced back to the earliest times.

In 1753 for example, a John Blackburn (who also had a wife Mary) was mentioned in Whitby Parish registers. His occupation was listed as Master Mariner. Whitby records also show a John Blackburn, churchwarden at Whitby Parish Church in 1791.

Interestingly on 17 January 1822 the Whitby registers record the birth of James Watson Blackburn, son of James Blackburn (master mariner), and Isabella, at Stepney in London, on 20 July 1820. The child wasn't baptised until the following 17 January at Whitby.

In 1823, the same or another John Blackburn who owned pews seating ten people in Whitby Parish Church was absent in London, and was being represented by a Mr William Willis at a meeting concerning the use of the church pews. The registers previously show that a John Blackburn had a son John born 15 October 1757 who isn't recorded in Whitby until the following 17 January, when the Blackburn family returned to the town to have their child baptised.

The alternative version of Elizabeth's background, namely that she came from a wealthy background, is discussed in Besant's Book about the life of Captain Cook in which he reveals:

> *"I am indebted to the Rev. Canon Bennett of Shrewton, Wilts, for information respecting Elizabeth Batts which no one else now possesses. She belonged to a highly respectable middle-class family, connected with various manufactures and industries...For Cook to marry into so substantial and respectable a family marks a social lift corresponding to his promotion in the navy."*

Besant also professes that Elizabeth Batts' *"manners were good and full of dignity, and that she was well educated. She loved to tell how on the day of her wedding she walked with Mr. Cook across the meadows to the church. Therefore she was living outside the town of Barking. As her grandfather came originally from Essex, she was probably staying with relations. The newly-married pair went to live in Shadwell, where Mrs. Cook's mother, then Mrs. Blackburn, resided. Afterwards they removed to the Mile End Road."*

This contradiction to traditional stories about Elizabeth's poor background deserves further research but is beyond the scope of this book other than to say that Besant's own book reveals many interesting insights that contradict or add to the Cook family story, including the fact that James Cook junior had a small fortune of his own. As stated earlier it's easy to see that if James junior did desert his post and live his life out in North Yorkshire, his fortune would not be available to him as a 'dead man'. Besant confirms this fact: "*With her pension and her share of the profits of the books and with other things — such as the inheritance of her sailor son's fortune* [i.e. James Cook junior], *sworn under £5000 — Mrs. Cook became a wealthy woman."*

THE MYTH OF THE SOUTH SEA SHILLING

By tradition James Cook senior was sent at the age of sixteen (some say eighteen) by his father who was a Scottish migrant farmer living in Ayton, Yorkshire, to work for William Sanderson, who some say was the brother-in-law of Skottowe's wife. Sanderson was a haberdasher of Staithes, a small fishing village near Whitby in Yorkshire. Early stories (glossed over in some modern works) claim that Cook stole a south sea silver shilling from Sanderson's till and was sacked, though others assert it was all a misunderstanding as Cook had exchanged a bright silver one for an older one of his own. There are countless versions of this story, most based upon others and embellished in the telling.

Arthur Kitson (quoting Besant) in his book 'The Life of Captain James Cook' relates one version which is probably completely erroneous.

> "*... After painting Saunderson's character in colours of a rather disagreeable hue, as one too fond of his grog for himself and his stick for his apprentices ...Cook stole a shilling out of the till, packed up his luggage in a single pocket-handkerchief, ran away across the moors to Whitby, found a ship on the point of sailing, jumped on board, offered his services as cabin boy,* [and] *was at once accepted...*"

According to the Internet Baldwin Project

> "*When he was quite a little child; James was set to work on the farm of a Mr. Walker near Marton,* [interestingly Walker is the surname of Cook's Master in Whitby who employed him on his ships]. *The farmer's wife took an interest in the little boy, and perhaps taught him his letters, but he did not go to any school till later. When eight years of age, Cook's family moved to Great Ayton, a village about five miles from Marton.*"
>
> The page goes on to relate another version of the South Sea Shilling Myth:-
>
> "*A girl had come into the shop to buy some article, and she had paid for it with a very bright new coin,—what*

was then called a "South Sea; shilling." Cook was so struck by its beauty; that he wanted to have this shilling for himself and he took it, putting into the till in its place a shilling of his own. But it happened that his master had also noticed this bright coin, and when he found that Cook had got it, he was very angry, almost as angry as if the boy had stolen it. There was a great dispute. Cook thought that he had done no wrong, because he had put in the till another shilling in place of the new one. His master was so angry that he went to Cook's father about it. The end was that James was taken away from Mr. Sanderson's shop, and apprenticed at Whitby to a Mr. John Walker, a ship-owner. He did not—as has often been supposed—run away to sea."

This would appear to be just another tale. Surely shillings would come and go in the till many times during the course of a day's trading. Any later customer could have been given the bright shilling as change. Who knows what the true story is, but in reality it seems likely or at least a possibility that in reality Cook simply 'Took the King's Shilling", an expression that at that time meant someone who agreed to volunteer for the Royal Navy (or as an army recruit). Though Cook didn't actually join the Royal Navy immediately, he did so later, and it's highly likely that this common phrase was used to describe his entering a maritime career, particularly if he had by that time decided to take Royal Navy examinations.

HIDDEN OFFSPRING?

Perhaps the most perplexing mystery is that James Cook senior and his wife Elizabeth may have had more than the generally accepted six children.

We know that he and Elizabeth Batts were married by licence as Records tell us that on his return from a voyage on the Northumberland,

Cook returned to England in the autumn of 1762 and on 16 December he appeared personally and:

> *"...made an Oath that he is of the parish of St Paul Shadwell in the County of Middlesex – aged upwards of Twenty one Years and a Bachelor and alleged that he intended to intermarry with Elizabeth Batts of the parish of Barking in the County of Essex aged upwards of Twenty one Years and a Spinster and further alleged that she hath resided in the said parish of Barking for upwards of four Weeks last past not knowing or believing any Impediment by reason of any pre-contract, Consanguinity, Affinity or any other Cause whatsoever to hinder the said intended marriage and he prayed a Licence for them to be married in the parish Church of Barking aforesaid.* [Signed] *Jas. Cook".*

An extract from the register of St. Margaret's Church, Barking, Records their marriage:-

> *"James Cook of ye Parish of St. Paul, Shadwell, in ye County of Middlesex, Bachelor, and Elizabeth Batts, of ye parish of Barking in ye County of Essex, Spinster, were married in this Church by ye Archbishop of Canterbury's Licence, this 21st day of December, one thousand seven hundred and sixty-two, by George Downing, Vicar of Little Wakering, Essex."*

After their marriage Mr. and Mrs. Cook lived for a time in Shadwell and then made their home at 7 Assembly Row at Mile End, Stepney (later 88 Mile End Road) in the area now known as Tower Hamlets. The general consensus in modern times is that James fathered six children, though there are some intriguing possibilities that there were others.

THE CHILDREN OF JAMES (Senior) & ELIZABETH nee BATTS

(Conventional history lists the six offspring as follows)

JAMES or 'Jamie' (1763-1794)
Born: 13 October 1763, Shadwell, London
Baptised: St Paul's, Shadwell
Died: 25 January 1794 aged 30

James Cook Junior passed his examination for Lieutenant in May 1782 and was consequently appointed as Commander of the 16 gun sloop Spitfire, a Royal Navy Tisiphone-class fire ship that had been launched on 19 March 1782, having been built at the shipyard of Stephen Teague of Ipswich and Commissioned in March 1782 under her first Commander Robert Mostyn. At first, the ship carried out service in home waters, particularly in the Channel and served as a peacetime vessel following the end of the American War of Independence. After a series of refits she was re-classified as a 14-gun sloop-of-war for service in the war against the French and was involved in the capture of a sizable number of French privateers and small naval vessels. She also sailed to the British stations in North America and West Africa and joined in the Napoleonic Wars. In June 1793, the Spitfire's Commander, Philip Charles Durham stepped down. It was announced that his replacement would be James Cook, son of the (even then) famous James Cook RN

NATHANIEL or 'Nat' (1764-1780)
Born 14 December 1764
Died 5 October 1780

Like his father and elder brother, Nathaniel lost his life at sea when he was just sixteen and serving aboard the Man-o-war 'Thunderer' commanded by Commodore Robert Boyle-Walsingham (1736 – 1780), an Irish sailor and MP who had been born Robert Boyle, the younger son of Henry Boyle, 1st Earl of

Shannon. The Thunderer went down in the Caribbean during 'The Great Hurricane' (Huracán San Calixto) of 1780. The storm, which is considered the deadliest Atlantic hurricane on record, killed an estimated 20,000-30,000 people. Nathaniel was a midshipman at the time, a term that meant an officer cadet, having, according to some accounts, entered the naval Royal Academy at Portsmouth aged eleven. This has to be examined in the light of the comment by J.C. Beaglehole in his work 'Cook the Man' that relates that Nathaniel and his brother James (on paper) entered naval service at the age of five and six respectively so that they could 'earn time' to be able to claim a number of years' service as midshipmen and be able to take their lieutenant's examination at an early age. Beaglehole commented:-

> *"... In due course the certificate would be forthcoming. It was a conventional piece of chicanery in the navy; some distinguished men owed their early advancement to it."*

ELIZABETH or 'Elly' (1767-1771)
Born 1767
Died 9 April 1771

Little is known of Elizabeth and dates of her death vary widely. Most accounts say she died aged three whilst her father was at sea, but in the book *'Life of Captain Cook'* by Dr Kippis, which is also repeated in a letter to the Naval Chronicle on 7 February 1803, we are told she died aged five. In Captain Cook's wife's obituary in the Gentleman's Magazine 1835, she was described as Elizabeth Cook, a female child who died of Dropsy, (a former medical term for Oedema involving fluid retention in the body). This account gave the child's age as "about 12". Could this be a printer's error regarding the death of Elly who is generally recognised as dying in her fourth year, or was there another later unrecorded daughter also named Elizabeth?

JOSEPH (1768-1768)
Born 26 August 1768
Died 13 September 1768

It is said that Joseph died aged one month old on the day his father set sail for Australia. It's said that Cook never saw the baby.

GEORGE (1772-1772)
Born 8 July 1772.
Died Thursday 1 October 1772

George also died as a child at the Cook's home at Mile End. He is said to have been a sickly child and it is likely that James Cook did see this baby, because he didn't set off on his voyage until five days later.

HUGH or 'Benny' (1776-1793)
Born 22 May 1776
Died 21 December 1793

Hugh was born at 88 Mile End Road, London and was the Godson of Hugh Palliser, another famous 'Yorkshire Lad' from Kirk Deighton. Hugh Cook was named after his Godfather (a fact, according to the Gentleman's Magazine, that was confirmed by Elizabeth Cook herself) and was described as *'a tall fine youth'* whose ambition it was said, was to become a clergyman. We aren't told why he was nicknamed 'Benny', though he may have had a second Christian name. We are told that his mother had already purchased the advowson (appointment to a vacant ecclesiastical benefice or church living) in advance of his expected ordination, but he died 'unacquainted with the fact'.

Hugh's death from diphtheria or scarlet fever (accounts vary) seems to have been a genuine tragedy as confirmed by the Hon. Archivist of Christ's College Cambridge, Professor Geoffrey Thorndike Martin. In answer to a query as to whether any other students died at the college at the same time, he replied that:

> *"There seems to be no evidence of an epidemic: Hugh's contemporaries in College seem not to have died at a tender age."*

The Professor went on to provide the added information that in J. Peile's, Biographical Register of Christ's College, II (Cambridge, 1913), it is recorded that Hugh was born at Mile End, London, was schooled at Merchant Taylor's, and was admitted to Christ's College on 15 February 1793, aged seventeen. He died on 21 December the same year, and on 24 December was buried in the nave of Great St Andrew's. In his memory, his mother left £1000 to the parish in 1835, a huge sum of money for those days.

The above list accounts for six children born to the couple, which appears to be generally accepted and confirmed by historians, though most quote the authority of Beaglehole in his book 'Cook the Man' *"He had time to father six children, only one of whom survived into adulthood. At home at Mile End, in the little house with a garden where Boswell visited him..."*

There are however a number of credible records that suggest Captain James Cook RN and his wife Elizabeth had other children and as we have seen, that his son Captain James Cook junior also became a father. Intriguingly, the baptism registers of St Paul's Church, four miles from Shadwell have the following entries, but given that there are two more children named James amongst them and that Cook may have not been at home during those years, we must assume that all, or maybe some of the children belonged to a different James and Elizabeth Cook.

> *October 7th 1775. James son of James Cook by Elizabeth, his wife.*
>
> *September 4th 1778. Thomas son of James Cook by Elizabeth, his wife.*

October 29th 1779. James son of James Cook by Elizabeth, his wife.

July 30th 1782. John son of James Cook by Elizabeth, his wife.

June 15th 1783. Elizabeth Ann daughter of James Cook by Elizabeth, his wife.

CREDIBLE EVIDENCE

Without further research, linking any of these children with our own James and Elizabeth Cook amounts to nothing but pure speculation, however, more credible evidence does exist which points to the fact that James Cook (senior) and Elizabeth could actually have living descendants. Most of the claims of decendancy don't simply spring from spurious claims. Instead they are backed up by dependable evidence from separate families whose historical genealogical details often correspond. More to the point, they are repeatedly supported by information from other sources such as newspaper obituaries, press reports and even an Australian official town guide which directs visitors to the grave of one of Cook's descendants.

The earliest record in this respect dates from 1794 and provides seemingly incontrovertible and verifiable written confirmation that James Cook senior had two married daughters. The Gentleman's Magazine (Volume 75, Page 182, 1794) in its obituary for James junior, lists all of James & Elizabeth Cook's children including two married daughters. On the contrary, the York Herald of 21 May 1825 reported that their ONLY daughter had died in 1779. A report on the burial of James Cook (junior) tells us his sisters had naval connections.

...The captain, had he lived, bid fair to have been an ornament to his profession. His body was brought to

> *Cambridge, Feb 4, and buried in St Andrew's church, in the same grave with his brother, who died there of a fever, in his 18th year, Dec 28, 1793. One brother was drowned in the Thunderer, in a storm;* ***and two daughters were married to naval officers, who were both drowned.*** [My emphasis]"

The Salisbury Journal for 17 February 1794 repeated a story from a Cambridge newspaper published on 10 February which also mentioned the two married daughters; as did the Lloyd's Evening Post on Wednesday 12 February 1794. Unfortunately the newspaper reports don't make it clear whether it was the daughters themselves who drowned, or their naval officer husbands. Given the men's profession, one would be inclined to think that it was their husbands. Without names and dates, it is difficult to determine the truth of the matter but it is at least possible that the two daughters may themselves have been drowned, particularly as it's noticeable that there appears to be no mention of them in the extensive and generous will of Elizabeth Cook (with three codicils) dated 28 April 1833.

Enduring Myths

Before we dismiss the rumours and evidence concerning James junior's survival as mere speculation or pure fantasy, it may be pertinent to examine some other Cook family myths that have since been questioned, exposed or disproved by later investigative writers.

THE DEATH OF JAMES COOK SENIOR

Even the death of James Cook senior RN, the great circumnavigator, has achieved a mythical status. How he died and in what circumstances has both a 'received' version, and alternative accounts, depending when they were written and by whom. A correspondent of the London Morning Post for 3 November 1817 for example claimed to have come across an islander working as a ship's cook who was personally present among the natives on Hawaii at the time of the great navigator's murder. The newspaper gave the following first-hand account which differs entirely from the accepted historical account:-

> "...*Captain Cook who was in want of wood as well as water, had perceived near the shore an old hut, which appeared to him to be neglected and gone to decay; and the wood of which he thought to be drier than that of newly felled trees; he therefore gave orders to pull down the hut, without having first consulted the natives. Neither he nor his people, doubtless, knew (and after the turn the affair took none of them could learn) that*

the place was tabooed. The islanders did not hesitate a moment to prevent, by a desperate attack, an act which they considered as an impropriety. They killed some of the workmen and put the others to flight. Probably those who escaped did not know the real cause of the attack which was so fatal to a part of the crew. The Negro cook appeared much affected by the recollection of his abode in Owhyhee, and ardently desires to return thither."

THE ROYAL NAVY MYTH

Beaglehole in his biography of James Cook senior stated that the Captain was not above creating his own myths, including the fact that his two sons were entered as having served on his ships when they were not actually there. James junior and his brother Nathaniel can be found on the muster rolls of the Scorpion in 1771, where James senior is listed as Commander and his sons as 'Commander's servants' serving from 31 August to 29 November in that year.

To be fair to Cook, he was being no less duplicitous than other naval officers of the time who took part in this widespread 18th century deceit. Though in theory it would have been classed as a naval offence, the procedure was widely tolerated in order to give the children of naval Captains additional 'sailing time' on their official naval records. Consequently, these young men were able to pass their Lieutenant's exams much earlier and so achieve early promotion.

The Mariner's Mirror for November 1962 tells us that *"It is now obvious that Cook also made provision for his sons in this way"*. It also confirms that James junior appears in the muster roll for the Endeavour as a Second Lieutenant's Servant on 7 April 1769, replacing one A.B. William Harvey. The ship was at that time at sea, heading for Tahiti meaning that it was impossible for James junior to have joined the ship to take up his post. Despite this, he was still 'on the books' as a crew member when the ship returned to Deptford on the afternoon of 13 July 1771. James junior also undertook another 'ghost voyage' as an Able Seaman on the Resolution; we are told that the Muster Rolls reveal that he *'Entered on 01 August 1773 – Discharged on 30 April 1775 at Cape Town.'*.

Similarly, Nathaniel Cook entered on 19 August 1773 and was similarly discharged on 30 April 1775 at Cape Town. Nathaniel Cook also apparently appears as a 'ghost crewman' on the Resolutions muster roll for the 1776 voyage.

THE CLAPHAM MYTH

There was once a common conception that Captain Cook (senior) lived in Clapham, an error which is sometimes repeated in publications in modern times. A balcony known colloquially as 'Captain Cook's Quarterdeck' at the back of the building where he supposedly lived was allegedly haunted by James Cook's ghost. It was also believed that Captain Cook would regularly write his name on the window with his ghostly finger on each anniversary of his death in Hawaii. However, despite the name given to the building, James Cook senior never actually saw it. Mrs Cook *did* live there for a number of years along with her son Hugh though and it is said James junior visited the building when he was not at sea.

A letter now in the National Library of Australia, Canberra, (Nk 9528) from Elizabeth Honeychurch of Mile End, to a Mrs McAllister is dated 29 March 1792 and gives us a rare view into Mrs Cook's snobbish character. She appears to have held herself in high esteem after becoming the wife of a naval commander, a feature of her personality that alienated her old friends and neighbours somewhat. The letter also gives us an insight into the status of her son James:-

> *"...I can say but little for Mrs Cook who I think might have found time to have sent you a few lines tho' she lives in high Life at Clapham & keeps a Footman—her eldest*

> *son is a Lieutenant has a Horse and lives in stile*[sic]*; the youngest is design'd for a Clergyman, he is a fine tall youth".*

Another letter to Frances McAllister (née Wardle) of Philadelphia, who was Cook's cousin and who had stayed with Mrs Cook at the time of her husband's first voyage, reflects how Mrs Cook had become estranged from those who knew her: - *"...Mrs Cook has left Mile End, gone to live in Surrey, but where I cannot tell, nor neither do I much care..."*

No doubt Mrs Cook's high opinion of herself was dented when she was named in evidence given after the murder of Mrs Elizabeth Richards in 1823 according to a book with the long-winded title: *'Clapham, with its common and environs: containing an historical and topographical description of the parish and manor, principally extracted from parochial documents, with a catalogue of indigenous plants growing in the neighbourhood, list of population, rates, and other parochial occurrences, from 1603 to 1827, and a list of the inhabitants, corrected to 1841',* by David Batten.

In it we are told that Mrs Richards of Clapham Common was killed with a poker and that two sailors seen nearby were suspects. Though Elizabeth wasn't implicated in any way, the details are quite interesting with regards to the Cook family when we learn that Charles, the son of Charles Richardson who was mentioned in connection with the case, was Mrs Cook's servant at that time.

THE COTTON TREE MYTH

Though James Cook senior did live in London, he was away from home so often that Elizabeth, his wife must have lived the life of a virtual widow. As we have seen, the idea that he resided at Clapham in what became known as 'Cook's House' with its 'Cook's Quarterdeck' has no foundation whatsoever. As early as 1859, the myth that

Captain Cook senior had planted trees on Clapham Common when he lived there had already been debunked by local residents. At an evening lecture at the Clapham Literary and Scientific Institution on 15 April 1859, Henry Whitehead MA the Curate of Clapham gave the following address:-

> *"There is a species of poplar on the Common, some very graceful specimens of which stand opposite Mr. Parrott's house a little to the North of the Parish Church. Of late years they have suffered considerable injury, having lost many of their top branches in gales of wind. They are beautiful even now, but I am told that about twenty years back they presented so picturesque an appearance that a gentleman, well known for his aesthetic taste, was in the habit of driving down from London once a week and stopping his carriage near the Polygon for the express purpose of admiring them. They appear to be of a kind that very quickly decays. People about here call them the cotton trees, and are under the impression that they were first brought here from the islands of the Pacific by Captain Cook. There is even a tradition that the great tree (another species of poplar) once known as the Seat Tree, so called from the seats which were formerly round it, was brought here by that celebrated navigator. I should be sorry to believe that the tradition that Captain Cook had any connection with Clapham has no solid foundation. At all events Mr. Wright, in whose possession the house known as Cook's House now is, assures me that he never lived there, but that it was occupied after his death by his widow, who died at a very advanced age about twenty-four years ago, of whom I doubt not that many here present this evening have distinct recollections. Her memory is perpetuated in this parish by her bequest, known as Cook's gift, to poor widows."*

J W Grover, who gave a local history lecture in 1885 entitled 'Old Clapham' confirmed that the person who really planted the rare trees on the Common was

"...the great navigator's eldest son also known by the name of Captain James Cook, RN. The seat tree stood near the present site of the temperance fountain. It was destroyed by a gale in 1893, and its surviving stump replaced some years later by the large black poplar which is just to the west of the fountain."

THE FREEMASONRY MYTH

Some internet websites have claimed that James Cook senior was a Freemason, a point referred to by The Grand Lodge of British Columbia and Yukon who tell us that *"As a matter of interest it is believed that in 1778, Captain James Cook became the first Freemason to set foot in what is now the province of British Columbia."* If so, this has relevance to his son James junior who would be a 'Lewis' (son of a Freemason) and as such could have joined a lodge at the age of eighteen, rather than twenty one, giving us yet another possible source of information which might enlighten us about the hidden history of this remarkable young man. Unfortunately, the claims that Cook senior was initiated into Lodge of Industry No 186 seem to fall down when we discover that the lodge was only given its warrant on 15 January 1788, nine years after Cook died. There is however the possibility that James Cook junior was the Captain Cook in question.

The history of the lodge tells us that its members originally met at the Black Friar in Playhouse Yard in Blackfriars but changed its venue over the years until 1808, when its warrant was suspended apparently because its members had admitted a black man as a member. Some accounts claim that the man was a freed slave, which would breach the rules that only 'free-born' men could be admitted as a Freemason.

For those who might wish to research Masonic documents regarding the possibility that James junior was a Freemason, the meeting places of the lodge are recorded in the lodge returns in Freemasons' Hall in London.

It was first constituted on 15 January 1788 at the Black Friar in Playhouse Yard in Blackfriars. By 1789 its home was the George and Crown, Broad Street, Bloomsbury; and in 1791 it was at the Duke's Head, Phoenix Alley, Long Acre. The following year, 1792, the lodge was to be found at the Black Horse in Boswell Court, Carey Street but by 1793 had moved to Phoenix Court, Long Acre as well as at the Grapes in Wardour Street Soho and the Sun, Great Windmill Street Haymarket. Later in 1794, the year that Cook junior died, the Bull and Ram in Old Street became its home. An enquiry was made to Mr P Aitkenhead, Assistant Librarian at the Library & Museum of Freemasonry who kindly checked all of the membership registers for the Grand Lodge of England prior to 1800. He found that there were no entries that contained the name of James Cook, so it would seem then that the belief that James Cook was a Freemason is an erroneous one, unless of course he became a Freemason overseas in a freemason's lodge that was not under the control of the Grand Lodge of England.

The Marriage of James Cook Junior

We return to the question of whether Captain James Cook junior was married. It's been said that James Cook junior married at either Dunnington, Huntingdon or Huntington; all near York and the couple had at least one child. Though Robert appears to be the most usual name mentioned for the child, rumours also claim that James later had a daughter, Rebecca, who married John Barnard Smith. If this is correct, the question remains; did the births take place before or after his supposed 'death'? Finding the marriage registers poses many difficulties. If the supposed wedding was before James junior's 'death' then where would it have taken place? There appears to be no matching entries extant in the York district. There is however a marriage registered on 26 December 1851 at Witham, Essex between John Barnard Smith, son of Edward Smith (aged twenty seven – born circa 1824) and Rebecca Cook (aged twenty four – born c1827). Though the bride's father's name is given as John Cook.

Stories have persisted through the years in North Yorkshire that James junior originally married a servant girl and that his mother was extremely unhappy at the match as she considered his bride to be too low class for him. Elizabeth Cook is said to have refused to recognise the 'low class' woman as her daughter in law and we know, following James junior's reported death, that his mother is widely reported to have actively destroyed correspondence and items belonging to James junior and his father. Why?

Elizabeth's close friend John Leach Bennett wrote at the time of her death:-*"...few memorials of Captain Cook came into my hands as executor of his widow. In fact for a few years before her death Mrs Cook had employed herself in destroying letters and papers..."*

One wonders if the destruction of the letters and other papers was a sign of abject grief, or was she attempting to destroy forever all correspondence that might serve as evidence of the circumstances pertaining to her son's supposed drowning. More to the point, might she also have consciously destroyed any documentary evidence pertinent to James' wife and offspring, so as to ensure that they wouldn't have any means of making a claim upon the fortune of her son and herself?

Another question we might ask is if the story of his marriage is true, could James's wife be the mysterious lady who he is said to have written to on the night of his supposed death, the mysterious Mary mentioned earlier, *James Cook the younger, gent., and Mary his wife, of the Manor of Balsden* or someone else entirely?

The 'London Public Advertiser or Political Literal Diary' on 29 January 1794 (and other publications) mentioned the mysterious lady:-

> *"Yesterday about noon the body of a sea officer in a Master and Commander's uniform, was found on the beach near Warden Lodge on the Isle of Wight; and by a letter dated Friday evening, 7 o'clock, from Pool,* [sic] *to a Lady at Portsmouth (per favour of Captain Cook, and other circumstances), there is every reason to believe he is the unfortunate person"*

Not having a surname for the mysterious lady does make it almost impossible to find out who she was, though we might speculate that she could have been another wife or perhaps a mistress who was living somewhere in Poole. The woman could of course also have been his mother, but why should she be in Poole? It is curious that her name was not mentioned in the press so did the newspapers actually know the woman's real name but for some reason choose not to make it public. Intriguingly there is another entry for a James Cook marrying in St Mary's, Portsea, near the naval centre of Portsmouth in 1791. The entry (number 469 – page 469) reads:-

> *"James COOK bachelor & Ann WORM spinster both of this parish (Portsea St Mary, Hampshire) married in this church by banns 8 May 1791 – X groom made his mark & bride signed in the presence of John WARN & Richd GUDGE."*

The fact that the groom made his mark rather than signing his name does seem to discount the possibility that this was the literate son of James Cook senior. Though if we return to the rumours that James Cook junior married in a village near York and later had a daughter, there is perhaps a more intriguing avenue to follow, could he have married twice or even bigamously? Though this theory is of course wild speculation, there does appear to be some deep reason why James Cook was described as *extremely melancholy* when he left to join the sailors from the *Spitfire* to return to their ship on that fateful night. Another marriage we must consider is the one recorded in the city of York itself. This particular wedding was recorded on the 24th June 1794, (after the supposed drowning),

Some Claimants to Descendancy

We need to question the fact that if conventional history insists that there are no direct descendants of the renowned circumnavigator Captain James Cook, why do prominent biographers continue to perpetrate this fact without question. It would seem that given the profound lack of public information regarding James junior and his death, most choose to follow the writings of biographer J.C. Beaglehole, from which all modern works about Cook's family appear to draw their information. Most writers tell us that the only descendants of the Cook family descend from Cook senior's married sisters, Margaret Fleck and Grace Cocker.

This again brings us back to claims to the contrary and old rumours that continued to circulate in North Yorkshire that James Cook junior didn't really die, and that his so called 'drowning' provided a convenient smoke-screen which allowed him to desert his post and return to his family roots in the rural surroundings of Sheriff Hutton area, near Malton. If true, then direct descendants of the famous Captain James Cook (senior) are almost certainly alive today.

Though unbelievable to most historians, there are a number of people around the world who sincerely believe they are descended from both Captain Cooks (senior and junior). What follows is a short summary of some of their claims. No doubt there are many others.

SIR ERNEST SHACKLETON (Antarctic explorer)

Shackleton is said to have 'masterly unravelled' the case of the marriage of James Cook the younger, proving that certain documents from Admiralty papers, especially those written just before Cook junior's alleged death and other documents from when he was commander of the *Spitfire*, bore James junior's signature which matched those on letters in his descendants' possession. All of them, he said, were identical and written on the *Spitfire* and elsewhere between the years 1789 and 1794. He also stated that Frank Wild, (who claimed to be his first cousin), immediately on his return from one of his early Antarctic expeditions, received a medal from Edward VII, who specifically referred to Wild's kinship with Captain Cook. In addition, King George V also mentioned the same family relationship a few years later.

COMMANDER FRANK WILD CBE, RNVR, FRGS.

Arguably the most credible claim to a direct decendancy from James Cook senior comes from Commander Francis 'Frank' Wild the Antarctic explorer who lived from 1873 to 1939. Wild was born in Skelton, North Yorkshire, only a short distance from Marton, where James Cook senior brought up his family.

The New Zealand Evening Post in 1928 published a report about the fact that the Hawaiian authorities had invited 'Commander Frank Wild, second in command of the late Sir Ernest Shackleton's South Pole Expedition' to attend tho sesqui-centenary of the discovery of Hawaii because of his lineal descent from Captain Cook. Wild at that time was farming in Northern Zululand, on his cotton plantation 'Quest Estate', named after Shackleton's vessel. Wild was far from wealthy and was doing railway work to 'keep the pot boiling.' The rather long articles that were published about the festival sought to clarify whether Cook really had any direct descendants, but are very useful to researchers in naming those that said he did. Wild himself always claimed, *"I have always believed that my mother, born Mary Cook, was a direct descendant of the explorer."*

Frank Wild was born John Robert Francis Wild and was the eldest of eight sons and three daughters born to Benjamin and Mary Wild (daughter of Robert Cook of West Lilling, near Malton) Robert believed he was the grandson of Captain James Cook senior

Benjamin Wild was born in Newcastle and was a schoolteacher by trade. Mary was born in Lilling near York and was a sewing mistress. We are told she was also convinced that she was a direct descendant of Captain James Cook senior and that her father Robert Cook was a descendant of the great explorer.

Until relatively recent times, James Cook junior's survival was said to have been common knowledge in this area, a belief that was also shared by some older families in nearby Whitby, which also has Captain Cook connections.

The story that has passed down through generations virtually without change, relates that in order to draw suspicion away from himself after his desertion, James changed his name to either John, George or Robert (stories vary) and returned to the village, possibly moving to Coatham near Redcar at a later stage under his real name in order to be in the same area as his grandfather (Captain Cook senior's father). At a time when communication was lax in such relatively isolated country areas, a change of Christian name would probably have been sufficient to allow him to successfully avoid the authorities, despite local people in the Sheriff Hutton district being aware of his real identity. It has also been suggested that the authorities may even have known where he was but were willing to ignore the fact so as not to stir up a hornet's nest, following the official line regarding his 'burial', in many senses of the word.

In making the claim, Frank Wild would have had nothing to gain from revealing his mother's lineage. What's more his reputation would certainly have been at risk if he had made spurious claims about being descended from James Cook RN, particularly as his assertion is based on descent from a man who supposedly drowned without marrying or having any children.

Wild's claim was supported by a Miss Norah Ward of Sheriff Hutton (near Malton Yorkshire) who died in 1981. She too claimed descent from the younger James Cook, who she said had married at nearby

Dunnington, She was the granddaughter, she claimed, of Robert Cook through her maternal line. Miss Ward firmly believed that James deserted from the navy and came to live in the area, initially next door to her mother's cousin who remembered a man coming to Robert Cook's house when she was a child. The man claimed to be the boy's father and assured him that he was 'born in wedlock'; even though at that time she didn't understand the phrase. That man was the allegedly 'dead' James Cook junior who asked if he could stay at Robert's house. His request was refused point blank by the boy's mother based on the fact that James had deserted Robert and his mother years before. James later went on, it is said, to farm at Cornborough and / or West Lilling, both places close to Sheriff Hutton.

In 1835 Sheriff Hutton was described as comprising "*Lillings-Ambo, Sheriff-Hutton, and Stittenham, and containing 1278 inhabitants, of which number, 756 are in the township of Sheriff-Hutton*". It gives the populations of Cornborough as 63; Farlington 170, Lillings Ambo 208 and Stittenham as 81.

We can see that even at that time the population of the farming community was quite large, though it was spread over a number of separate close-knit communities, many of which would be inter-related by marriage. Given this fact, it would at least be possible for those who knew who James Cook junior was, to be able to keep the fact of his 'resurrection' a secret from the wider world. Even so, if it was a secret it must have been a fairly open one, otherwise the legend could not have endured within the region and as far afield as Whitby where James Cook senior spent his early sailing days.

The marriage of the younger James Cook, if it took place after his 'death', would have presented an even bigger problem and it is difficult to see how he could have married legally without the naval authorities tracing him, if indeed they wanted to. Those who believe the marriage happened claim that he used a different first name to throw any enquirers off the scent.

Family Search records on the internet show that a George Cook married Elizabeth Flint at Dunnington near York on 16 February

1790 (prior to Cook junior's death) and there is an interesting record showing that a Robert Cook, son of Elizabeth Cook (no father mentioned) was christened at Rillington cum Scampston on 8 December 1796. Rillington is another nearby rural village situated not far from Sheriff Hutton.

Though we can't verify that the man who called at Robert's home was definitely the supposedly drowned Captain Cook junior, we can positively confirm that Robert existed. He appears in the 1851 census as a sixty year old farmer at Lillings Ambo with his wife Mary (née Hutchinson) and seven children, namely Isaac, William, Hannah, Elizabeth, Robert, Jabez and Mary. It was the youngest child, Mary who later married Benjamin Wild and gave birth to Frank Wild the arctic explorer.

MRS NORTHWOOD (Wild's sister, living in England)

When the authorities in Honolulu were planning to celebrate the sesqui-centenary of Cook's discovery of Hawaii they began a search for a descendant of Cook to attend the celebrations. Frank Wild's sister corresponded with them and repeated the family claim:

> *"Captain Cook is our great-great grandfather. My mother's father was Robert Cook, of Lilling, York. He was the son of Captain Cook's son, who was married at Huntington Church, near York. They had one son, Robert Cook, born 1791. He married Mary Hutchinson, and they had ten children, of whom the youngest was Mary. She married Benjamin Wild, of whom the eldest son was Frank, now Commander Wild."*

It is interesting that Mrs Northwood mentions the place of marriage as Huntington near York, not Dunnington, though this could be purely an error of memory or due to the intricacies of parish boundaries. Another witness, Marie Von Holt of London claimed she was descended from Thomas Brown, one of the first planters in Hawaii, who was associated with officers who served with Cook and their descendants. She verified to the Hawaiian authorities that from what she had been told she fully believed Mrs. Northwood's claim to be a direct descendant of Captain Cook.

MISS M. A. COOK (Pembroke, Ontario, Canada)

Curiously, Miss Cook's story matches closely with that previously given by Mrs Northwood.

> *"I am his great-great granddaughter. My father, William Cook, was born on August 1, 1836, and died in Canada September 24, 1806. His father, Robert Cook, was baptised on November 24, 1790; died May 17, 1881; buried at Sheriff Hutton, being the son of James Cook, who was drowned January 25, 1794, aged 31.* [He was actually 30 but described as being in his 31st year]. *Robert Cook was the only grandson of Captain Cook*".

The validity of this claim appears to be more credible because Miss Cook appears to have heard that James Cook had supposedly drowned yet she still claims to be descended from him, confirming in her view that any marriage in Yorkshire must have taken place before the date of his supposed drowning.

FLORENCE M. WOLTER (Australia)

It's intriguing that Australia, New Zealand and Canada also have families that have publicly claimed a direct descent from James Cook senior. The Randwick City (Australia) tourism website in 2015 unapologetically pointed its readers to the grave of one of Cook's Great grand-daughters:-

> *"Turning left into Frances Street you can walk through St Jude's Cemetery, the resting place of Florence M. Wolter, great grand-daughter of Captain James Cook, the English explorer credited with having discovered Australia."* The now defunct website can be viewed on the Wayback Machine internet archives or on the modern Randwick City Tourism Site.

ELIZABETH CHAD (Port Colborne, Canada)

In a similar vein, Elizabeth Chad of Port Colborne in Ontario, Canada made the following claim on a New Brunswick website in 2015, namely that one of Cook's daughters, Ann, lived to be ninety four.

> *"I am the great granddaughter of Captain James Cook and I thoroughly enjoyed my walk into his life and that of* [the ship] *Marco Polo. I am going to visit Newfoundland next year and I will search for additional information. I am named after his daughter Elizabeth she lived to be 94 but she was only 10 when Captain Clarke* [referring to James Clarke Captain of the Marco Polo] *passed away."*

CHARLES H. CLARK (of Yorkshire U.K.)

Charles Clark wrote two letters saying that the mothers of Commander Wild and himself were sisters. He appeared to have known his grandfather, Robert Cook, who died in 1881, saying:-

> *"My Grandfather, Robert Cook, believed himself to be the grandson of Captain Cook. There was an old book with the name James Cook written on it. It was given to Sir Ernest Shackleton, and this signature was examined*

> *by an expert to see if it agreed with naval documents. Shackleton would not have supported this claim if there had been any doubt".*

He adds that Wild was received by King Edward and King George as a descendant of Captain Cook, and was also received as such in New Zealand and elsewhere.

SUSANNAH ROWE (of Woodbridge, Suffolk)

A death notice in the Ipswich Journal, for 25 March 1843 ran as follows:-

> *"19th instant at Woodbridge aged 78, Susannah the beloved wife of George Rowe Esq. retired surgeon of the 69th Foot and daughter of the late Captain James Cook, Royal Navy."*

Though this is not a direct claim to descent, the fact that she was born in 1765, two years after James Cook junior, would make her a likely candidate for decendancy.

ANN RUMSEY (of Colchester, Essex)

Reference to another claimant as one of Captain Cook senior's unrecorded children occurred as death notice printed in the Malton Messenger newspaper (Yorkshire) on 28 September 1867:-

> *"DEATH of CAPTAIN COOK'S DAUGHTER*
> *The death is announced, in the parish of St. Martin, Colchester, of Mrs. Ann RUMSEY, widow, in her 104th year. It is an interesting circumstance that she was the daughter of the celebrated circumnavigator, Captain Cook, who was*

> *massacred by the natives of Owhyhee, in the South Sea Islands, and that she was born only a few years after the accession of George III to the throne of England"*

The story is given some credibility in that it was repeated in other newspapers not only in Britain but also overseas; but could she really be another unknown daughter? If she was indeed Cook Senior's daughter and died aged 103, this would make her year of birth either 1763 or 1764. The woman, who lived on Back Queen Street in Kidderminster was described as 'of independent means'. Further research however, appears to suggest that this Ann may be one and the same as Ann Cocker, a daughter of Christiana Cocker, James Cook senior's sister, who married a Mr Tree. Christiana in turn had a daughter Ann who married William Rumsey. This Ann died in 1867.

MR PETER LAWSON (Ricall, near York, England)

Despite having no verified evidence, Mr Lawson states:- *"Our family is one of those with a passed down story of being related to Captain Cook...My great- great grandfather was a John Cook 1820 to 1894 (married to Jane Hutchinson 1827/1915) and his father was a William Cook 1788/1859."*

Interestingly, the surname Hutchinson, occurs both in the testimony of Mrs Northwood, which was mentioned earlier, and elsewhere in the text

Mrs BARCLAY (New Zealand)

The Ohinemuri (New Zealand) Regional History Journal of September 2011 reported:

> *'Over the past 12 months a small team of volunteers have been planning to erect a large six-ton ship's anchor,*

donated by the Royal New Zealand Navy, on the site of where Captain Cook is thought to have landed during his historic voyage up the Waihou River on November 21, 1769...A most interesting feature of the Kopu ceremony was the fact that of those who stood in the breeze to watch the unveiling and hear the speeches was married to a descendant of Captain Cook. He was Mr W. Barclay of the Public Works Department, Paeroa. Mrs Barclay, who was prevented by illness from attending, was a direct descendant of the famous 18th century navigator. Her family in England—named Adams and living in Ealing, London—had in their possession Cook's personal seal. Mr Barclay said he had seen this seal often before he came out to New Zealand, little thinking how the association with Cook would recur in years to come'.

KEN FULKS (Canada)

Ken Fulks, a Canadian storekeeper, arrived at Townsville airport, Australia in October 1964 at the end of a tour of Cook related landmarks, claiming he was the great-great-great-great grandson of Captain James Cook via a direct line from Cook's granddaughter, Elizabeth.

The Cook Family of North Yorkshire

If we accept that there is a possibility that James Cook junior did have a wife, even a common law wife, in Yorkshire, let us examine again what real evidence there is to prove he fathered a child or children.

A letter is held by the New Zealand National Archives which was sent by Jabez Atkinson, a cousin of Frank Wild, to a C R H Taylor (written on 15 May 1939). It presumably concerned the Cook inheritance and mentioned photographs which apparently have not survived.

Interestingly Mr Atkinson appeared to believe the drowning story:-

> *"Dear Sir, thanks for your letter dated May 10th. In reply I do not wish to make any official claim,* [presumably to the Cook fortune] *there are so many of cousins of mine on the male side (in Yorkshire especially).... The photo is a photo of Jabez Cook of Lilling Yorks, a son of Robert Cook of Lilling who was a son of James Cook, who was a son of the Captain Cook".*
>
> Jacob Atkinson goes on to say... "*Captain Cook came from the working class. When he had made a name, he married a Parson's daughter, a woman far above him socially (you would have to live in England to understand the caste system). Mrs Cook refused to associate with her husband's relatives. Mrs Cook refused to acknowledge the wife of her son James Cook as she was a servant (domestic)*
>
> *James Cook* [junior] *was drowned (probably caused partly through drink) as a result of this my grand-father*

Robert Cook, was a lifelong Teetotaller, a very unusual thing in those days.

James Cook acknowledged my grand-father, but my grand-father always preferred his mother. My grandfather visited Mrs Cook [Elizabeth the wife of Captain Cook senior], *but Mrs Cook was cool and did not wish to acknowledge him. A Minnie Cook in Canada, daughter of Robert Cook, (son of R Cook of Lilling) is the real claimant (if living). My cousin Sir Frank Wild, son of Annie Wild (Annie Cook) was acknowledged as a descendant of Captain Cook by two kings of England, King Edward and King George. Shackleton of South Pole fame proved that the signature in the ship's log of James Cook corresponded with the one in a bible in my grand-father's possession.*

Legal proof or no, ***there are perhaps hundreds of descendants on the male line.***

Yours truly, J Atkinson. [my emphasis].

Though the question of who Minnie precisely was must for the time being remain a mystery, Wikipedia explains: *"As a first name, Minnie is a feminine given name. It can be a diminutive (hypocorism) of Minerva, Winifred, Wilhelmina, Hermione, Mary, Clementine and Amelia".* Census returns may eventually reveal her full name, just as they can help us to identify individuals in the Cook family who were once living in and around the area of Lilling in North Yorkshire, the area where it is said Cook junior fathered one or more children.

The Cook Family Genealogy

It is remarkable that a James Cook, born in exactly the same year as James Cook junior, the 'drowned' naval Commander, was to be found as a pensioner in an almshouse at nearby Kirkleatham. Could this be where the mysterious 'drowned' mariner eventually ended his days?

The following gives a full detailed account of the Cook family in North Yorkshire in earlier times as described in the census. Those not interested in the genealogical connections may wish to skip this section though it does provide vital information for those wishing to research Cook's family connections further.

ROBERT COOK (The suspected son of James Cook junior.)
1841. Robert, a farmer, was aged 50 (born circa 1791) and living at Lillings Ambo West. With him was his wife Mary aged 40 (born circa 1801) and children John 8; Isaac 7; George 5; William 4; Hannah 3; and Elizabeth 2.

1851. Robert and family were still living at Lillings Ambo, West Lilling. He was then described as being aged sixty and a farmer of 56 acres. This census says he was born in Huntington near York, where one account says James junior was married, and that his wife Mary, aged 49, was born at Sheriff Hutton in around 1802. The same census reveals that all of the children were born at Lilling. Those recorded at the address in 1851 were Isaac, William, Hannah, Elizabeth, Robert, Jabez and Mary. The last three were marked as Scholars with Mary being described as a 'Scholar at home'.

1861. The census reveals that the family were still living at Lillings Ambo. Robert was described as a farmer of 50 acres who was born in Dunnington (again where James Cook junior is said to have married in another version of the story). Dunnington is approximately five miles from Huntington which was declared as his birthplace in the previous census. Mary, his wife and children, Isaac, Jabez and Mary were still at home. By this time Robert's daughter Hannah had found employment as a house servant to Joseph Richardson and his wife Ann. Richardson was an agent for the Calder and Hebble Navigation Company and their family were living in the Calder and Hebble Navigation shop on Bridge Street in Wakefield. Robert's daughter Elizabeth, now aged twenty one, was a servant in the house of Henry Thompson, a corn merchant in Nethergate, Nafferton Yorkshire. Young Robert, aged nineteen was an apprentice carpenter & joiner to Joseph Geary in the village of Claxton.

1871. Robert, now aged 80, was still farming at West Lilling (40 acres). He confirmed his place of birth as Dunnington. Still at home was his wife Mary and son Jabez, together with Robert T Cook, Robert's twelve year old grandson, a scholar who was born in Marske near Redcar. Robert T was probably the son of the thirty eight year old John Cook (born Lilling) who was to be found in the census at Marske where he was an Ironstone miner and living at Cowles buildings with his thirty one year old wife Hannah. She was born in what appears to be Radwell Bedfordshire. Their children were Sarah E, a scholar aged 10 born in Marske; Jane A, aged 8 born at Windgate, (Wingate?) Durham; Henry aged 6 born in Redcar; Hannah M, aged 4, born in Marske; and Eliza, aged 2, also born in Marske. Hannah, wife of John Cook, was living at 18 Dale Street Marske with her children. She was described as Head of Household and married (i.e. not a widow).

1881. Robert T was 22 and a blacksmith by trade. Sarah E was a dressmaker. Jane A had become a schoolteacher and Henry was an apprentice joiner.

It's interesting to see that the towns of Redcar and nearby Marske repeatedly appear in the census details, as they are closely connected to the Cook family. James Cook senior's sisters lived there and their father went to live there in his retirement and was eventually buried there. Remarkably, Robert Cook was still alive aged 90 in 1881, as was his wife Mary, now 80. Robert is still described as a farmer employing one man at West Lilling. Jabez his son (farmer) is still at home; as is granddaughter Elizabeth aged 12, a scholar born in Stockton, Durham.

1901. We now find Jabez aged 57 married and farming at West Lilling. His wife Ruth who was a year younger was born at Rillington. Their daughter in law Eunice Shields aged 21 (born Sheriff Hutton) was staying with them on census night.

1911. Jabez was 67 in the census and was still at West Lilling. Because of the additional details on the 1911 census form, we now learn that he is not only a farmer but also an employer and that he and Ruth have been married fifteen years but they never had children together. Because son in law John Shields (a horse man on the farm) and Eunice Shields (described as daughter in law in the previous census but now described as an unmarried daughter), are in the household, we could speculate that they are Ruth's son and daughter by a previous marriage. Another anomaly that is hard to explain are the three children living with them, all described as scholars. They are Natham (not Nathan), Henry Harrison 9, John Harrison 7, and Ruth Harrison 4. Perhaps they are Eunice's children from a previous relationship. Though the house or farm is not named, the census tells us that it has six dwelling rooms not including scullery, toilets, work rooms and other ancillary rooms.

CHRISTOPHER COOK

1851. There were three dwellings, possibly farms, between the homes of Robert and Christopher Cook, the middle one being uninhabited.

Christopher Cook was aged 57 and born at Lilling circa 1794. He was also a farmer, of 104 acres. Christopher's wife was Hannah aged 41, who was born circa 1810 at Whenby between Sheriff Hutton and Brandsby, Yorkshire. Their children (all scholars born in Lilling) were George 14; Sarah Ann 12; Thomas 11; Mary 9; Joseph 7 and Jane 4. They apparently had two employees living in; Mary Barker a house servant aged 15 (born Sheriff Hutton circa 1836) and William Cundall a farm servant aged 18 (born 1843).

1861. Christopher, now aged 68, was retired and living at Settrington near Malton in a dwelling situated four houses along from the White Bear Inn. He is described as a retired farmer aged 68 living with wife Hannah and children George, Joseph and Jane. His other children don't seem to appear on the census for that year and may have died or left home by that time. George went on to become a self-employed boot and shoemaker living at 5 Daubney Street, Cleethorpes and appears in a later census at the age of 65 when he was living with his wife Harriet, aged 64, who was born at Warter, East Yorkshire.

JOHN COOK

1851. John, aged 17, was born at Lilling circa 1834 and was a farm servant in the household of Thomas Consitt in Strensall near York at the time of the census. Given his age John was probably the eldest son of Robert and Mary.

1881. His children were (presumably) Hannah, Eliza and William, aged 9. Interestingly, William was born in Marske. All children were described as scholars. Hannah was head of household (i.e. John was absent). John had evidently changed his profession because in 1891 John and Hannah were to be found at 29 Gilmore Street, Thornaby running a grocery shop. Children Hannah M, Eliza and William, now grown up, were still at home. The two girls were school teachers whilst William had become a clerk.

ANN COOK

1851. Ann was possibly the youngest child of Robert or Christopher. Though she was born in Lilling, in 1851 she was staying at the home of her grandparents James & Catherine Waite in Burythorpe near Malton.

SARAH A COOK

1871. Sarah was aged 26 and had been born in Lilling circa 1845, though now living at 14 New Bridge Street, Micklegate York in the home of her aunt Tabitha (aged 43 born at Sutton on Forest) and uncle Joseph Dresser, a pork butcher aged 42 (or 45). He was born at Whenby near Sheriff Hutton. Sarah may have been acting as a nanny to their children Thomas James 8, and Mary Ann aged 6, both of whom were born at Whenby, a village between Sheriff Hutton and Brandsby.

N.B.:- The above information has been drawn from a number of sources and though it is believed to be correct, no guarantees are given. Genealogists and researchers should verify the facts for themselves to ensure complete accuracy.

The Official Enquiry

Returning to the actual maritime incident that resulted in an official Royal Navy enquiry; the evidence given on oath by Captain Watson and the sailors who were there at the time, proved inconclusive due to conflicting evidence.

It should be remembered that lying whilst giving evidence under oath was a very serious matter, so much so that prior to 23 July 1812, the penalty for lying under oath to the Surveyors General in matters concerning 'Laws for collecting His Majesty's Revenue in Great Britain' was punishable by death. No doubt the men giving evidence in this particular enquiry were fully aware of the consequences of bearing false witness.

With regards to Watson's character, it emerges from the inquiry documents that he was not well liked by many of his men. We know that he was lying to William Turner, the shopkeeper, when he said that Cook had hailed him from the Spitfire's boat and that he had personally thrown him a rope. There also appeared to be some sort of unmentioned stain on his character from a previous incident. The Surveyors General in charge of the enquiry remarked that they had been unaware that Watson had been previously either dismissed or suspended from the Revenue service by an order of 22 June 1781.

The Surveyors General, Edward Burrow and Arthur Hammond heard the official enquiry. Burrow was born on 21 November 1726, the first son of Joseph Burrow, collector of customs at Whitehaven in Cumbria and became collector of customs at Kingston-Upon-Hull in about 1756. He took up a similar position in Glasgow in

1774 and became a Surveyor General of Customs in circa 1790. Hammond had previously been a magistrate prior to his appointment as a Surveyor General. He died at Southampton in 1807 at the age of sixty seven.

The confusing evidence that was presented to the inquiry so perplexed the two men that on 27 March 1794 they concluded their first report and proposed to their superiors that a second inquiry should be conducted where each mariner would be cross examined individually in order to get to the bottom of the mystery.

Extraordinarily, as stated previously, the full names of the other sailors who accompanied Cook in the Spitfire's small boat were never revealed at that enquiry, nor do they seem to appear anywhere else afterwards. This situation beggars belief, particularly as seven men and a coxswain were said to be aboard the vessel that night. Surely someone, somewhere, knew exactly who these men were and if so why were these mariners never named in the investigation? In normal circumstances, friends and family would have been desperate for an explanation regarding the loss of their loved ones. In some cases these individuals must have been the last people to see these individuals alive before they set off in the open boat to join the Spitfire. As such they would surely have proved key witnesses and could perhaps have given valuable evidence to the inquiry.

By the time the second part of the investigation took place it was plain that some sailors had reconsidered their positions with regards to what evidence they were about to give. There appears to have been a fair amount of collusion as to what evidence should be given. Some appear to have been in support (or in fear of) Watson, whilst others, it seemed, were determined for whatever reason to paint as bad a picture as they could of him. Some mariners even verbalised their common dislike of Watson and voiced their opinion that his actions at the time meant that he should be 'turned out', i.e. sacked. One went so far as to declare that in his opinion Watson should be hung for what he did!

THE WITNESSES

George Edward Dodd spoke of the fact that some men had said that they wished they knew anything that would get Watson turned out. Whilst giving evidence Dodd never once mentioned the conversation alleged to have taken place by John Carter the Chief Mate of the Grey Hound, or that he himself had uttered words to the effect that *"Captain Watson deserves to be hanged and I know it"*, a fact noted by the examiners.

William Harrington admitted that whilst in the cutter on its way to London to attend the enquiry, he had said something to the effect that in his opinion,Captain Watson's conduct on the night of the loss of the Spitfire's boat merited that Watson should be turned-out; and more importantly, it was still his opinion.

A**aron Twiststrap** was a Grey Hound crew member who swore on oath that he had wished it was in his power to turn Captain Watson out because he thought that Watson did not pay enough attention to the Spitfire's boat and that the sailors aboard it might have been saved if he had done so.

J. Harris There is some confusion with the name of this witness due to the use of John Harris, James Harris and Joseph Harris in these documents. Whether these refer to a single person or different individuals is unclear. It was Harris whom Dodd said was of the same opinion as himself, but he denied on oath that he had said anything of the kind. A similar denial was made by other sailors. It would seem however that some of the sailors had changed either their opinions or evidence for personal or career reasons or perhaps for fear of retribution in the event that Watson was eventually found not guilty of the charges against him.

Thomas Bagg, a.k.a. Baggs supported the Captain, and swore on oath that he had heard John Carter utter the words, *' I am afraid the Old*

Above left: An unusual portrait of James Cook R.N. senior (1728–1779). It is said to portray "a remarkable likeness…".

Above right: The apparently snobbish Elizabeth Cook. Nee Batts (1742–1835). Her family origins are still disputed.

Above left: A portrait often said to be of a young James Cook (senior), but which some researchers believe to be the supposedly drowned James Cook (junior).

Above right: Hugh or 'Benny', brother of James Cook junior who died aged 17 from either diphtheria or scarlet fever (accounts vary) whilst attending Christ's College Cambridge.

Above: Nathaniel Cook, another son of James & Elizabeth perished aboard the *Thunderer* in the Caribbean during 'The Great Hurricane' (Huracán San Calixto) of 1780.

Left: The grave slab marking the resting place Of Elizabeth Cook and her two sons Hugh and James in the aisle of St Andrew's the Great, Cambridge.

IN MEMORY
of CAPTAIN JAMES COOK, of the Royal Navy,
one of the most celebrated Navigators, that this
or former Ages can boast of; who was killed by
the Natives of Owyhee, in the Pacific Ocean, on the
14th Day of February, 1779; in the 51st Year of his Age.

Of Mr NATHANIEL COOK, who was lost with the
Thunderer Man of War, Captain Boyle Walsingham,
in a most dreadful Hurricane, in October, 1780;
aged 16 Years.

Of Mr HUGH COOK, of Christ's College, Cambridge,
who died on the 21st of December, 1793; aged 17 Years.

Of JAMES COOK, Esq; Commander in the Royal Navy,
who lost his Life on the 25th of January, 1794; in
going from Poole, to the Spitfire Sloop of War, which
he commanded; in the 31st Year of his Age.

Of ELIZ^TH COOK, who died April 9th 1771, Aged 4 Years.
JOSEPH COOK, who died Sept. 13th 1768, Aged 1 Month.
GEORGE COOK, who died Oct. 1st 1772, Aged 4 Months.

All Children of the first mentioned CAPT. JAMES COOK by
ELIZABETH COOK, who survived her Husband 56 Years, &
departed this life 13th May 1835, at her residence Clapham Surrey
in the 94th Year of her Age. Her remains are deposited
with those of her Sons JAMES & HUGH,
in the middle Aisle of this Church.

Right: A memorial plaque on the church wall that refers to the family grave in the Middle Aisle of St. Andrew's church. It contains added genealogical information.

Below: The old St Paul's Parish Church, Shadwell where James Cook junior was baptised. It was traditionally known as the Church of Sea Captains.

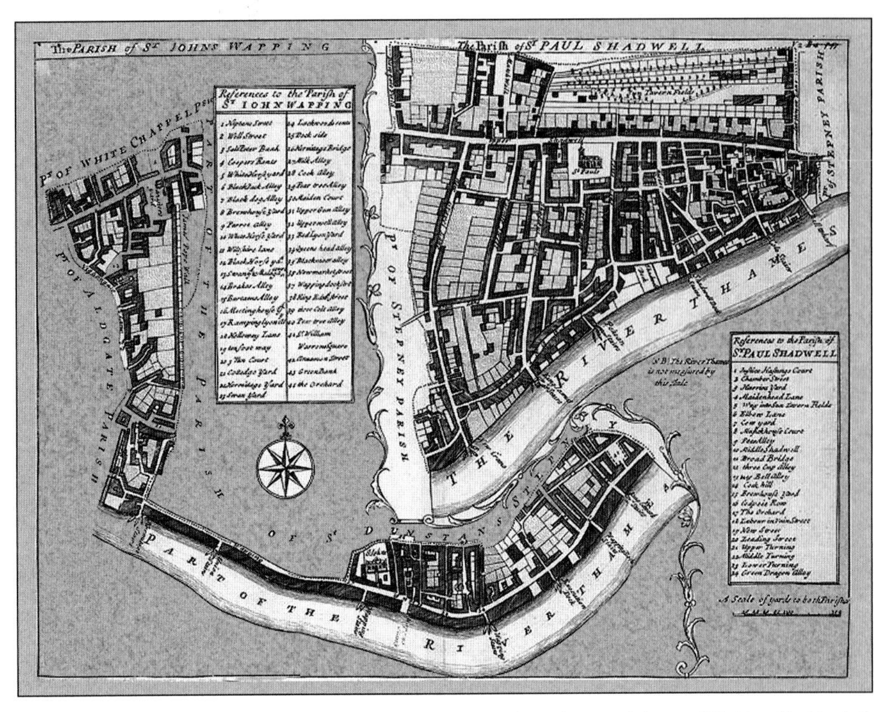

An 18th century map of Shadwell, London showing the position of St Paul's Parish Church (top right).

The historically accepted circumstances of James Cook's death in Hawaii were challenged by an islander working as a ship's cook (see 'Enduring Myths' in text).

A press cutting from the Salisbury Journal, 3rd February 1794, giving a report of James junior's death and perhaps the 1st mention of the mysterious farmer White.

Poole, Jan. 31. Captain Cook, of his Majeſty's ſhip Spitfire, was unfortunately drowned on Friday evening laſt, in attempting to reach his ſhip, which lay at anchor near the Bar. The Captain had that day taken a farewell dinner with ſome of h's friends, and was remarkably ſerious and melancholy; he left his company immediately after tea, being intent on ſleeping on board, as the ſhip was under ſailing orders—it blew freſh from S. W. and was very dark; he was adviſed to deſiſt from attempting a paſſage that night;—but his ſenſe of duty prompted him to perſevere, and he periſhed in conſequence.

Captain Cook was only thirty years of age, and was the eldeſt and only ſurviving ſon of the late celebrated navigator, and had he lived, he bid fair to be an ornament to his profeſſion.

The boat was ſuppoſed to have miſſed the ſhip in the night, and afterwards to have upſet on one of the ledges of rocks near the Iſle of Wight, as his body was found the next day near Yarmouth. The boat had drifted in cloſe to the place where the body was picked up; but no part of the boat's crew, which conſiſted of the coxſwain and ſeven men, have yet been found.

The above melancholy tidings, Capt. Roberts was deſired by the Admiralty to convey to Mrs. Cook, (the mother) who was nearly bereft of her ſenſes on the occaſion, having not a month ſince loſt her youngeſt ſon at College.

By the death of Capt. Cook, his Majeſty has loſt an experienced officer and loyal ſubject; his country, a warm partizan; his intimates, an active friend, and agreeable companion; his inferior officers and ſeamen, more than imagination can paint, or expreſſion do juſtice to. Not an enemy remains, for it was impoſſible he could ever make one.

The officers and ſeamen of the Spitfire have returned thanks to Mr. White, of Yarmouth, for his attention to the body. This benevolent farmer had made a very decent preparation for its interment at their arrival.

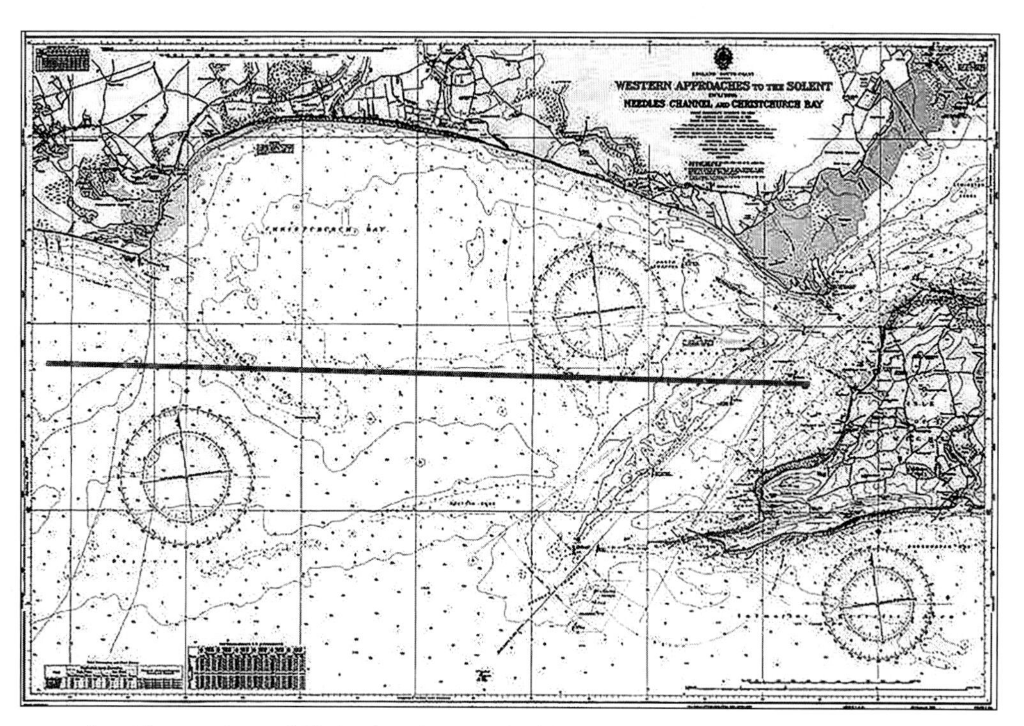

An old navy chart of Christchurch Bay. The black line shows the direction in which the wrecked boat carrying James junior would have travelled (left to right).

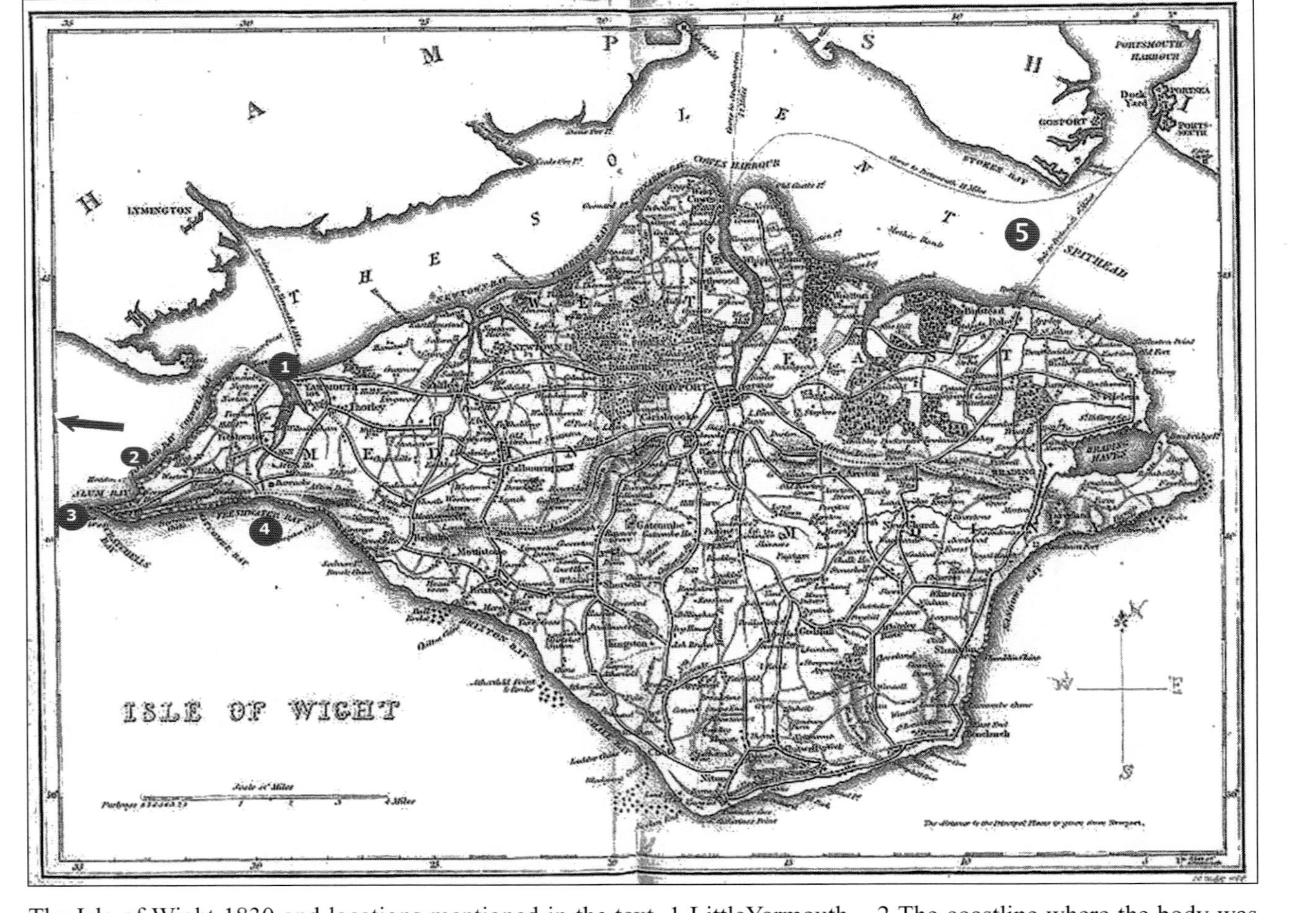

The Isle of Wight 1830 and locations mentioned in the text. 1 LittleYarmouth – 2 The coastline where the body was found – 3 The Needles – 4 Freshwater Bay – 5 Spithead.

98 Meteorological Diaries for January and February 1794.

METEOROLOGICAL TABLE for February, 1794.

THE

Gentleman's Magazine:

For FEBRUARY, 1794.

BEING THE SECOND NUMBER OF VOL. LXIV. PART I.

23	W brisk	29,73	47	7–10 rain, several showers of hail
24	W brisk	60	43	*7–10 blue sky, white clouds, stormy, snow
25	NE stormy	28,23	39	snow till P. M. tempestuous
26	N brisk	29, 7	36	sleet at intervals, snowy evening
27	N calm	28,96	37	frost, little snow
28	W brisk	29,10	38	thaws all day, freezes at night
29	SW calm	33	42	frost, snow, showers, thaw
30	SW moderate	37	41	clear sky, slight frost, thaws
31	S moderate	35	46	dark sky, tempestuous at night

The Gentleman's Magazine February 1794 confirms the stormy weather conditions on the night James Cook junior set sail to join his ship.

A typical revenue cutter of the time also named HMS *Greyhound*. It was renamed *Viper* in 1781.

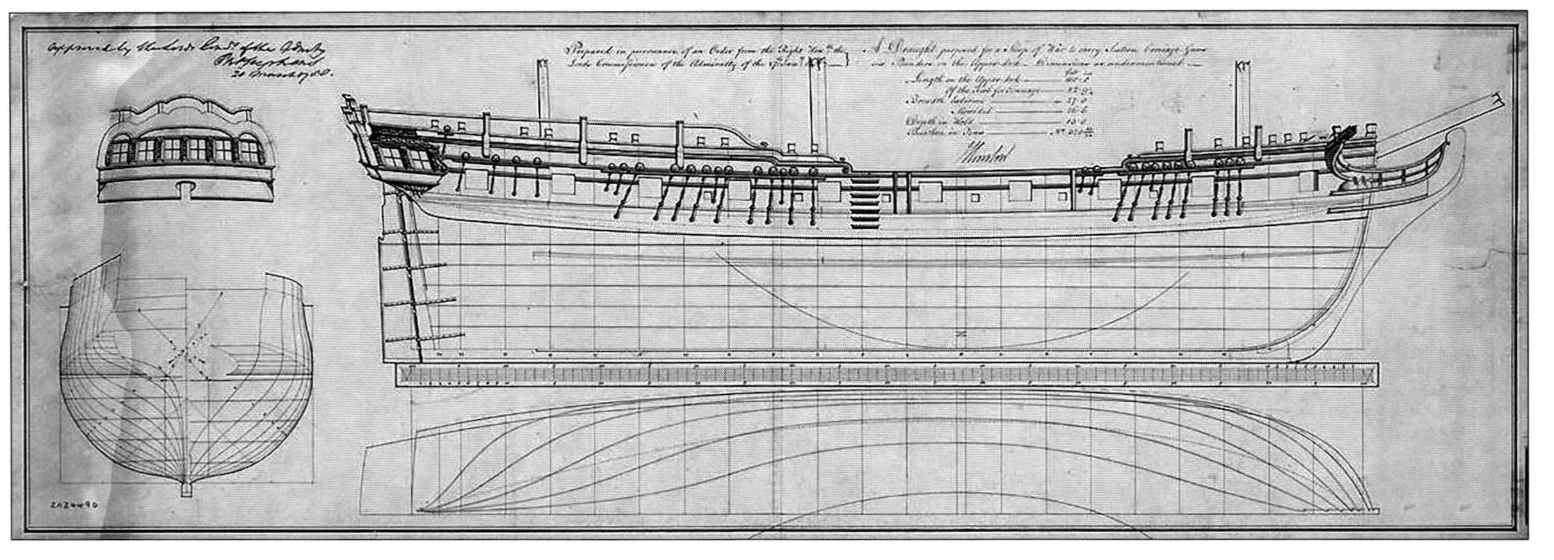

The building plan for the Excise vessel *Hound* whose crew accused Captain Watson of the Greyhound of causing the death of James Cook junior and his crew.

Spice Island, the seedy area of Portsmouth docks on Portsmouth Point where Cook's death was discussed in the grocery shop of his friend William Turner.

Above left: Captain (later Admiral) John Bazely. One of the two naval officers present in William Turner's grocery shop soon after news of young Cook's death was received.

Above right: Admiral Bowyer, who when in Turner's shop controversially said that if officers wanted to drown themselves they had a right, but shouldn't risk their own men's lives.

It is ſaid, upon authority which we believe may be credited, that an examination of a very ſerious aſpect to the parties concerned, has been ſome time in hand at the Admiralty, relative to the loſs of poor Capt. Cook and his boat's crew—when, if it appears aſſiſtance could have been afforded them, but was withheld from whatever motives it might be, it is to be hoped for humanity's ſake the proper perſons will be puniſhed as they deſerve.

A press cutting from the *St James Chronicle* newspaper, a.k.a. *British Evening Post* 13th March 1794 reporting the first indications of an Admiralty cover-up.

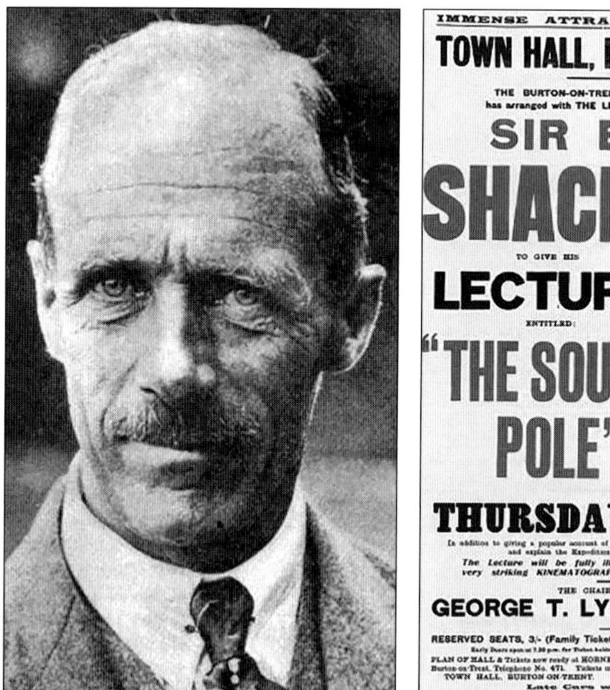

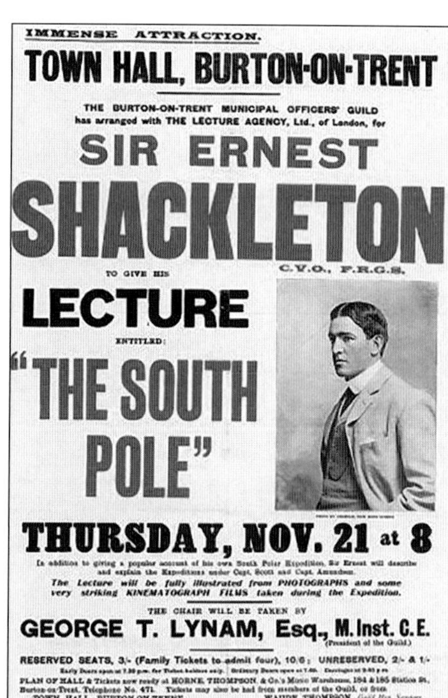

Above left: Antarctic explorer Frank Wild claimed direct descent from James Cook R.N. senior, via Mary Wild, (nee Cook) and her father Robert Cook of Lilling, Yorkshire.

Above right: Ernest Shackleton is said to have proved signatures belonging to James Cook jnr. after his death, were the same as those on Naval documents from his earlier life.

Worldwide claimants of direct descendancy from Cook senior, all claim connections with the same former residents of Sheriff Hutton, near Malton, Yorkshire.

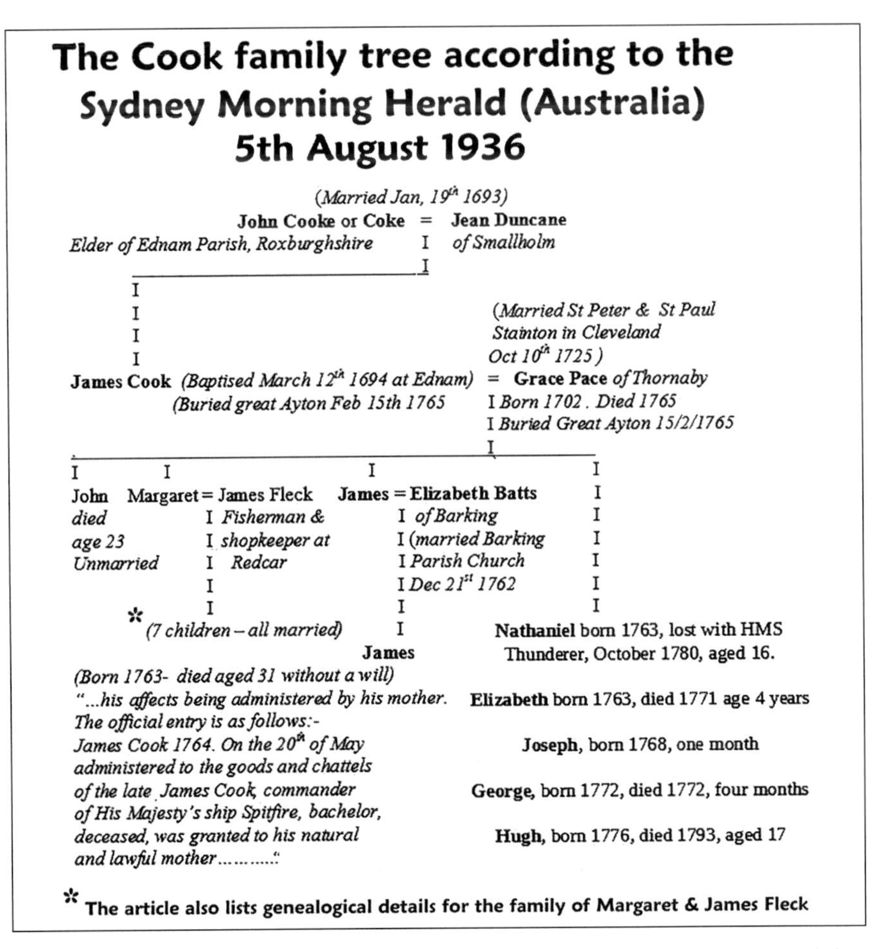

The Cook family tree according to the Sydney Morning Herald (Australia) 5th August 1936

(Married Jan, 19th 1693)
John Cooke or **Coke** = **Jean Duncane**
Elder of Ednam Parish, Roxburghshire — *of Smallholm*

(Married St Peter & St Paul Stainton in Cleveland Oct 10th 1725)
James Cook *(Baptised March 12th 1694 at Ednam)* *(Buried great Ayton Feb 15th 1765)* = **Grace Pace** *of Thornaby*
Born 1702. Died 1765
Buried Great Ayton 15/2/1765

John *died age 23 Unmarried*

Margaret = **James Fleck** *Fisherman & shopkeeper at Redcar*
(7 children – all married)

James = **Elizabeth Batts** *of Barking (married Barking Parish Church Dec 21st 1762*

James
(Born 1763- died aged 31 without a will)
"...his affects being administered by his mother. The official entry is as follows:- James Cook 1764. On the 20th of May administered to the goods and chattels of the late James Cook, commander of His Majesty's ship Spitfire, bachelor, deceased, was granted to his natural and lawful mother.........."

Nathaniel born 1763, lost with HMS Thunderer, October 1780, aged 16.

Elizabeth born 1763, died 1771 age 4 years

Joseph, born 1768, one month

George, born 1772, died 1772, four months

Hugh, born 1776, died 1793, aged 17

* The article also lists genealogical details for the family of Margaret & James Fleck

The Cook family tree according to the *Sydney Morning Herald* 5/8/1936 giving additional genealogical details. The newspaper also lists the Fleck family details.

Above left: The grave of Florence Wolter in Randwick, Australia with an inscription that states she is the Great Granddaughter of James Cook, 'discoverer' of Australia.

Above right: Sir Philip Charles Durham (later Admiral) 1763 – 1845 was Captain of the *Spitfire* prior to James Cook junior being appointed as Commander of the ship.

Turner's Almshouses., Kirkleatham, Redcar. Is this where the supposed drowned James Cook really ended his days? (See text Conclusion).

One of the Clapham 'cotton trees' said to have been planted by James Cook senior. They were in fact donated by his son, James Cook junior.

Captain Cook's Tree – Another of the trees planted around the churchyard at Clapham. It was later destroyed by lightning and replaced with a poplar tree.

WINDSOR CASTLE

18 June, 2010

Dear Mr Waters,

Thank you for your letter of 1 June enquiring about a letter sent to King George III reporting the death of Lieutenant James Cook in January 1794.

I have read through all of the King's surviving official correspondence in the Royal Archives for that month, but I regret that this does not include the letter in question. I am, therefore, unfortunately unable to provide you with a copy of the letter in which you are interested, and I am sorry to have to give you this disappointing reply.

Yours sincerely,

P. M. Clark

Miss Pamela Clark
Registrar, Royal Archives

The letter from the Royal Archives confirming that the supposed letter to King George III reporting James Cook junior's drowning was not in the collection.

An old photograph of the memorial to the killing of Capt. Cook senior at Kalakau Bay. (Source: Wikimedia Commons, Public Domain)

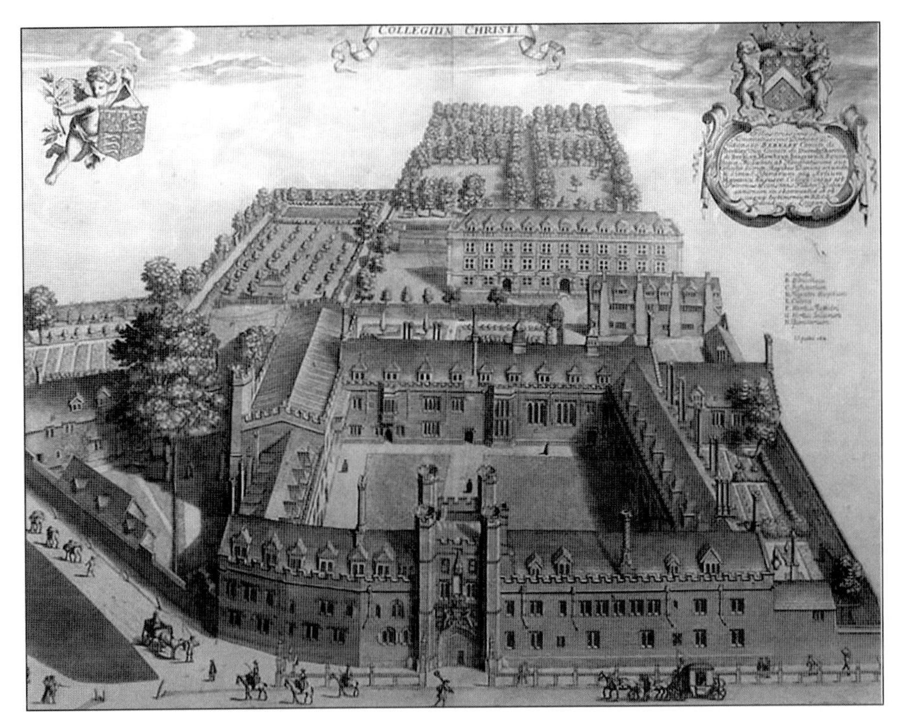

Christ's College Cambridge as it was in earlier times and where Hugh aka 'Benny' Cook studied when he died in 1793. (Public Domain Wikimedia)

Needles rock formation, (mentioned in the enquiry) as they were in the 18th century.

Left: An old painting showing the uniform of a Royal Navy Commander in the 18th century. (Public Domain - Wikimedia Commons)

Below: Totland Bay where James Cook junior's body was reportedly found,as it is today. (Source: Mypix, Creative Commons Attribution-Share Alike 4.0 International license)

Man [meaning Watson] *will sail with us again"*, and '*I do not think he will be turned out* [sacked] *this time unless you all keep in one story*'.

William Harrington and **Aaron Twiststrap** specifically denied they had ever heard Carter utter these words.

Stephen Watson (Captain of the Greyhound) wrote a long and rambling handwritten letter of defence, containing a number of spelling mistakes and underlined sections and signed in full with a signature (i.e. not with his mark). Spellings are as found:-

> *"To do away the hearsay tale as appears between Mr John Carter, Mate of the Greyhound and Mr John Butler Mate of the Hound, I must observe Captain Roberts in my charge says Butler accuses Carter of waving the conversation onboard the Hound when the death of Captain Cook was mentioned – If he waves there he continued to wave, for since he John Carter on oath says he told Mr Butler onboard the Hound that Henry Buck mariner of the Greyhound told him (Carter) that I Captain Watson in his (Buck's) opinion had not done my duty by paying that attention that I ought to have done to Captain Cook's boat + crew when alongside the Greyhound and that he, Henry Buck, was fearful what had become of her. – On this Mr John Carter + Mr John Butler are confronted + Butler on oath flatly denied overhearing Carter utter those words, adding that if he had, he should have enquired among the Hound's crew if any such discourse had drop't from the Greyhound's boat crew. – and what is still more extraordinary when the said Mr John Carter is confronted with Henry Buck from whom he says he had this news, he deny's recollecting that he ever said to Mr Butler that Henry Buck told him that I had not paid that attention I ought to have done to Capt. Cook + his boat crew which appears in book B page 42---The situation the Spitfire*

boat was in will prove the falsity of the charge respecting her bows being stove in and the fore thwart [A thwart being a strengthening strut or brace placed crosswise in a ship or boat that can also serve as a rower's seat] *being torn out of her, and the method in which she left the vessel is clearly proved on oath not to correspond with Captain Robert's account. Mr John Butler to close his examination on oath denys ever having told Capt. Roberts that my (Capt. Watson's) sailing from Spithead when the Spitfire came in was a proof in him, Mr Butler's, opinion of a guilty conscience in me. And as a very strong contradiction to the above assertion I beg leave to refer to the Hon. Commissioners + you gentlemen to Mr Turner's letter which is annexed to the report of this inquiry, proving my relating at Portsmouth to Admiral Bowyer, Captain Bagely and several other Naval officers in his presence the circumstances of Captain Cook's boat when alongside the Grayhound.*

That there has been and still is a conspiracy among the crew of the Greyhound to get me turned out appears to me very clear + certain, I must forcibly strike everyone from the following observation:-

When the general question (Book C No 301) is put to twenty mariners of the Greyhound, nine persons answered in the affirmative + declare they have frequently heard a general conversation viz 'That they wished they knew anything in their power to turn out Captain Watson that they might have none of the Sod' yet they answer they cannot fix on any one man that has discourse on this subject; this must strike the honourable board that it must have existed some time from the firmness by which they adhere to each other. George Edward Dod in Book C No 304 by oath accuses John Carter the Mate of saying I think Captain Watson ought to be hanged – and John

Carter when confronted on oath denys it; yet he recollects some part of the conversation then held with Dod.

Thomas Bagg mariner accuses Mr John Carter in Book C No 305 of being an adviser in their conspiracy of their crew and Carter denies the said accusations on oath. I shall not make any further observations on this subject but leave the matter to the observations and decisions of the Honourable Board.

Thirty three years I have faithfully at sea dedicated to the service of King & Country twenty eight of which I have been an officer under the Honorable Board of Customs in the different stations of Mate + Commander and I do flatter myself my exertions have ever been such as have rendered much benefit to the Revenue and to the coasting trade of this Kingdom. and in wartime, as will appear by the Privateers I captured in the last + present wars --- No officer in the Navy can in the most distant way accuse me of not paying attention to them when I knew they required it of me, on the contrary the numbers I have assisted and saved when in distress in the course of twenty-eight years are very great but at present so dispersed that I cannot immediately call them forward. – Two happen by chance to be on the spot, Lieutenant Bray. Late Commander of Nimble, Admiralty cutter and William Cranston, late Master of the Wasp, Sloop of War whose certificate I have delivered on the present enquiry for the inspection of the Honorable Board.

I am Gentlemen with the most respect, Your Humble Servant. Stephen Watson.

A postscript was later submitted as further evidence:-

Postscript

NB. I beg to state that I omitted the number of Privateers which I took in their Honors cruizers under my command

> *which amount to seven. And it is with further pleasure I can acquaint your Honors that in the space of 28 years that I never lost a man in all the boat duties during that time that were under my direction, nor any drown'd out of any of the vessels I had the charge of during that time. --- Which I hope will prove to their Honors in this length of time how attentive I must have been to this particular branch of my duty."*

There follows a summary, compiled by Watson, giving a selection of snippets from every mariner's evidence, with his own attached notes proving in his eyes how the evidence contradicted the charges against him.

SURVEYORS GENERAL CONCLUSION

The officials concluded that it was obvious from evidence given by crew members that the crew of the Spitfire's boat could have been saved at the time, but that:

> *"...Under the circumstances of such complicated evidence, where assertion and denial has been so often set against each other, it is impossible for us to discover with whom the false swearing lies. We can only observe that Mr Butler's evidence on oath does not fully correspond with the statement transmitted by Captain Roberts to the Admiralty, and he positively denied his having said that Captain Watson's leaving Spithead immediately on the arrival of the Spitfire was a mark of guilty conscience...."*

Those judging the inquiry in reference to the much repeated words said to have been uttered by Watson: *"Damn them, they are only man of war men and may go to Hell,"* said that had this been true it would have "*...evidenced not only great inattention, but wilful disregard to the boat on the part of Captain Watson*" and that in respect to Carter the Mate, "*...he is contradicted in several parts by Butler and others...and if he did not think his* [Watson's] *conduct wrong he seems to have failed in his duty to his superior officer in not appraising him of what had been said against him*".

In the concluding portion of their report, Burrow and Hammond determined that some crew members were hostile to their Captain but by the same token there was evidence that this was probably justified because he had even ignored the crew of his own patrol boat when they too got into difficulty that same night:-

> *"Your honours will perceive that upon the examination of John Miller No342, It does appear that he and several of the Grey Hound crew did express a wish that Captain Watson should be turned out and that he himself has said so because that being in one of the Grey Hound's boats that same night off of Great Haddon and in much distress, baling the water out of the boat with their hats, and flashing their pistols with gun powder as signals of distress, no notice was taken of them by Captain Watson's cutter which came near to them or within 20 yards, although they called out loud enough for them to hear, and hallo'd something back from the cutter which they did not understand, and the cutter was then standing for the land and left the boat in the same circumstances."*

The examiners also commented on the loss of Captain Cook's rowing boat when it could have been saved, declaring:-

> *"That he and his crew was once in safety and in the tow of the Greyhound's cutter after they were nearly 3 leagues* [approximately ten miles] *out of their way, nay directly*

the contrary way to Studland Bay, for they were far so seen and is perfectly clear. Why she was not so preserved in safety, whether her loss proceeded from harshness, obstinacy, fatigue or any other cause in Captain Cook and his crew, or for want of having that actual assistance afforded by the Greyhound, or that superintending care over her when in tow which the tempests of the night really commanded is a matter, all circumstances considered and the whole of the evidence in books A, B, & C, now delivered in, weighed and examined that must be, and is respectfully submitted to the wisdom, consideration & judgement of the Honorable board."

The document was signed by Burrow and Hammond and carried a P.S. *"Captain Watson having delivered as the inclosed letter from Portsmouth dated the 7th of April marked C, we annex it to this report."*

The document referred to was a somewhat fawning second letter of defence by Watson which appears to be a desperate last minute attempt to sway the Admiralty and persuade them of his innocence

"Greyhound Revenue Cutter - Deptford April 7th 1794.
Gentlemen,
Inclosed I beg leave to deliver my defence on the malice and ill-founded hearsay reports of Captain Roberts who charges me with the loss of the Spitfire's boat and crew, trusting from your wisdom and well known impartiality that justice will render me in your final decision thereon. I remain Gentlemen, Most respectfully your very humble servant, Stephen Watson"

His letter appears to add nothing to the evidence and continues to insist that Captain Robert's accusations were malicious and based on hearsay. A supporting letter from Captain Bray, who was in charge of another revenue cutter appears in the files, and appears to have accompanied Watson's own letter. It tells how when Bray was run down by a larger Dutch-built ship in his shallop revenue cutter 'Nimble', a shallop was a large heavy boat, usually having two masts and sometimes guns, Watson afforded him every assistance; and in the same year 1786, he rescued other vessels including a galley.

With regards to the allegations that some sailors had said that they wished to be rid of Captain Watson, it is clear that those hearing the witness's statements couldn't tell which witnesses were telling the truth.

> *"...eleven declared they had not heard any such thing pass, and nine admitted under several variations in their evidence that something like the words, others that the very words, some that disrespectful words passed against the Captain & others that he deserved to be turned out, that they would get him out if it laid in their power, some assigning as reasons for not taking care of Capt. Cook's boat, and others assigned no reasons at all'.*

They also concluded that perjury and lying had been committed by some sailors because assertions and denials were *"...so often repeated that there must be false swearing on the one side or the other, but on which we are sorry that it is not in our power to discover".*

All in all, the costly and lengthy inquiry seemed to have served no real purpose and appears to have concluded the matter without any proper resolution. I have been unable to find any reference to Watson being punished in any way for his actions.

(A full transcript of the inquiry which took the form of a legal trial and giving personal details of the witnesses (e.g. age, background, years in service etc.) is printed in the Appendix)

Author's Conclusion

It is easy to see that if James Cook junior actually deserted his post, the government would have been in a dilemma, given that he was a respected and well known personality of the time, not to mention that he was the son of an even more famous father. Acknowledging his death and even organising a mock burial would perhaps have been the easiest way to bring the matter to a close quickly. It does seem strange that virtually no records seem to have survived of the actual funeral and that the man's name has been all but obliterated from all naval records despite his high status. Wiping James Cook junior from the record books would be an effective way for the authorities to ensure future generations would never find out the circumstances that had proved an overwhelming embarrassment both for the Royal Navy, and his conceited, snobbish and well-connected mother.

We might well ask what sort of argument, situation, or sense of unhappiness would make a young man of thirty give up a promising naval career. We may never know, but it seems he arrived virtually penniless and depressed at the door of his son in North Yorkshire, assuring him that he was not illegitimate and seemingly begging for forgiveness from the boy's mother. Evading the authorities may have become easier over the years and if old rumours are to be believed, he decided to take up farming in the Sheriff Hutton area where the family had connections apparently settling into a rural lifestyle.

As stated previously, the Cook family also had connections with Coatham and Redcar where we know that James Cook RN's father went in order to retire from farming and live with his daughter Margaret, the wife of the fisherman James Fleck. It would seem that James junior never received the full amount of the bequest made to him in his father's will. Similarly, the bequest to 'James Cook of Redcar' (James Cook senior's father) was never paid either as he died in March 1779, ten months before news of his own son's death in Hawaii reached London.

The preponderance of people named James Cook is hard to untangle, particularly if a change of Christian name was later involved, even temporarily. Further research would be needed to discover if this is the same James Cook mentioned in North Yorkshire registers who registered the birth of a son (also James) at East Coatham on 27 December 1795, and seemingly another son William, on 12 November 1797? This James was described as a labourer. Further sons including Thomas were registered on 3 May 1800.

It would be easy to dismiss the existence of this North Yorkshire based James Cook as a convenient coincidence. However an entry occurs in the census for 1841 that lists a James Cook of exactly the right age to be the one in question. He was living in Turners Hospital at Kirkleatham, an almshouse close to Guisborough, around 50 miles from West Lilling. The almshouse is also in the same limited geographic area where James Cook senior's father and sister once lived. Sir William Turners Hospital was founded in 1676 and still stands today, not as a true hospital in medical terms, but as a charity-based alms-house or hospice with the aim of caring for twenty old men and women and twenty orphan boys and girls.

The 1841 census entries for the hospital contain a list of old men who are all given the occupation 'brother'. Intriguingly amongst them is one James Cook aged 78 who was born in Yorkshire in the significant year of 1763, the same year that James Cook junior was born. Even more interesting is that what appears to be a record of the same James Cook's death. According to the Superintendent Registrar' records for Guisborough Vol. 24 – page 229 he is listed as dying between April and June 1844.

Appendix

Full Transcript of the Enquiry by the Surveyors General

A full transcript of the 18th Century enquiry into the death of James Cook Junior. And the charges against Captain Stephen Watson of the coastguard cutter Greyhound

The report is contained in three bound documents marked *'BOOK A, B & C'* constituting a full record of the evidence brought before the Surveyor's General. Held over a number of days it provides revealing key details regarding the circumstances leading up to the supposed death of James Cook junior.

The somewhat long-winded introductory transcript of the first day is followed by easier to read entries of actual witnesses statements under cross examination. Together they present a rare and interesting insight into official naval proceedings of the time. Entries appear not to be in the actual words of those being examined (some of whom may well have been illiterate) but are rather a summary of their statements which were read out to them before being signed

Surnames and spellings vary throughout the document and there are a number of words struck out and amended, presumably when the entry was questioned by the witness before being signed by him. In the full transcript that follows, spellings are as found and surnames have been printed in bold type to help identify the individuals concerned. Like the explanatory notes in square brackets they are provided to aid the reader's understanding and do not appear in the original documents.

BOOK A Containing the evidence before the surveyors General of the Customs upon the investigation of Captain Watson's charge.

Transcript of entry for Friday 7 March 1794

"Surveyor Generals' Office – Friday March 7th 1794.
Present ***Mr Burrow & Mr Hammond****, Surveyor Generals.*

On a charge against ***Stephen Watson*** *Commander of the Grey Hound Cutter burthen 200 tons in the service of the Revenue of Customs, who upon proceeding to investigate the matter of the said charge, call in Mr* ***John Butler****, Mate of the Hound Cutter in this service, and the said charge, and* ***Captain Watson's*** *answer thereto being read in the presence of both parties, the following examination of the said* ***William Butler*** *is taken upon oath; in pursuance of the Honorable Board's directions of the 6th March 1794.*

This deponent [a deponent being an affidavit given under oath, or the person giving such evidence] *saith that on Monday the 27th of January last, being at anchor at Spithead on board the Hound Cutter of which he is Chief Mate and had then the command in the absence of* ***Captain Hawkins****, who was upon leave from the Honorable Board, and riding near to the Grey Hound Cutter belonging to the Customs, Mr* ***Carter*** *the Chief Mate of the Grey Hound with some of the crew came in his boat on board the Hound Cutter, whose names or number he does not know, nor should he know the persons agaïn if the whole crew of that vessel were now present, having had his conversation with any of them but said* ***Carter****, the Chief Mate who said to this deponent upon his coming down into the cabin 'what a sad accident has happened', to which this deponent replied ' what is the matter?' and he thereupon said 'Poor Captain Cook of the Spitfire Fire Ship was drowned', but he did not proceed any further to give this deponent an answer of any circumstances or other matters relative to the said Captain's death; nor does this deponent recollect that he* *at that time did ask the said Carter, how, or in what manner, the said accident happened; nor had this deponent either the curiosity or presence of mind to ask the said Carter who continued in the cabin along with this deponent nearly two hours (after drinking some grog together) any further questions nor did any conversation whatsoever pass between them, relating to the death of Capt. Cook and thereupon the said Carter, with his boat's crew rowed aboard their own Cutter which was then at anchor at about the distance from them of about quarter of a mile.*

And this deponent further saith that after the said boat and the crew had left the Hound, ***James Paine Bundell****, one of the deputed mariners on board the Hound, came aft to this deponent (who believes he was then upon the deck) and told him what the Grey Hound's people that had just left the vessel had been telling to the said* ***Bundell*** *and several of the crew belonging to the Hound, some circumstances respecting the death of Capt. Cook; which this deponent states upon his oath to be as follows*

'That they were cruising with the Grey Hound without the Pool bar; and blowing very strong, the Spitfire's boat came to the cutter, and asked them to give them a tow to their ship in Studland Bay; at the same time one of the sweeps that was lying over the cutter's stern, by her pitching, fell as they supposed on the boat and broke in. [sweeps were very long heavyweight oars used as an alternate source of propulsion for small sailing vessels when there was no wind.] *That the man that was heaving the lead laid it down and threw a rope to the boat.* ***Captain Watson*** *at the same time came upon deck and looking over the topsail told them to get forward by saying it was nothing but a Man of War's boat, and they might go to Hell, and they should let go the sail, by that the cutter had so much way, it tore away the thwart,* [a beam running laterally across the structure of a boat] *as they supposed the rope was fast to, on which they cryed Lord Have Mercy on them, and they never saw them afterwards. And this deponent is the more particular in the foregoing assessment, because he speaks from a written memorandum of his own, which he made on*

board the Hound on the 1st day of March instant, when he examined the said ***Bundell****, and also that* ***Storey*** *and* ***William Naylor****, two of the mariners belonging to the Hound Cutter, and to whom after he had made said memorandum, he read it over twice to each of the said three persons separately; and they said it was true and they were ready to Depone* [testify] *thereto upon oath.*

And this despondent further saith, that upon the 28th of January last in the evening, Mr ***Farris,*** *the mate of the Rose Cutter in the service of the customs came on board the Hound, as she was at anchor at Spithead, & related a story much to the same purport, as the Grey Hound's people had told on board the Hound; and further added that he had orders from* ***Captain Hurry or Harry*** *of the Navy residing at Yarmouth in the Isle of Wight; and another officer whose name he then mentioned, but the despondent does not recollect, to call on board the Spitfire (as the cutter was going to Spithead) and to give the same account to the officers on board the Spitfire; and despondent for then saith that in the conversation passed betwixt him and the Mate of the Rose Cutter, he understood that the account then related by the said Mate (and which the despondent has stated to be much the same purpose as the Grey Hound's people had told on board the Hound) was given by the crew of the Grey Hound Cutter to some of the crew of the Rose but did not mention the time when that passed; therefore their despondent doth not know whether it was given on board the Rose or not.*

And this deponent further saith that the day after he had received the first information from ***Bundell*** *but had not then had the same confirmed by* ***Storey*** *and* ***Nailor*** *(two others of his crew whom he has before mentioned) He met* ***Captain Watson*** *upon the Point at Portsmouth, where they civilly accosted each other, but this deponent did not then mention or give the least hint to* ***Capt. Watson*** *of what he had heard, or since.*

And lastly, this Deponent saith that none of the circumstances he has before stated arise from his own particular knowledge, but the account he has given, was such only as he became possessed of from hearsay and unless he had heard the said story and told that he would not have thought it worth his while to have mentioned it, even to ***Captain Roberts*** *when he sent for him; as he never even mentioned it to his Colleagues and Court at Shoreham.*

Taken & Sworn this 7th March 1794 before us – Adjourned till Tuesday 11th Instant.

The document is signed by Butler and countersigned by **Burrow & Hammond,**

Transcript of entry for Tuesday 11 March 1794

Tuesday 11th March 1794
Present Mr ***Burrow****, Mr* ***Hammond*** *and Mr* ***Barker***

[**James Paine Bundal,** a deputed mariner, meaning a sailor who had received a deputation to act as a Customs officer and who would be paid between £3 and £5 per lunar month, gave his evidence under oath and was asked a number of specific questions. His answer to each question was personally signed by him]

James Paine Bundal *aged 25 years a Deputed Mariner on board the Hound Cutter in the service of the Customs, being produced and sworn as a witness to support the charge in the presence of the accuser and the accused, answered upon oath to the following interrogations;*

1. *Give an account of what passed on the Hound Cutter, on or about the 26th January last betwixt you and any of the crew of the Grey Hound boat that came on board your vessel, relative to the death of Cat. Cook, and where was your cutter then at anchor, and where was the Grey Hound.*

   Both vessels were at anchor at Spithead. I cannot speak to the precise day but I recollect that in the month of January last, the Grey Hound's boat came on board with several of her crew under the Command of Mr ***Carter****, the Mate who was immediately upon* *his coming on board went down into the cabin to Mr* **Butler** *our Mate, and of the rest of the crew, some of them went down into the galley, and some of them remained upon deck, but the number of each that did so I do not know, nor even the number of sailors that came in the boat. Hereupon I went down into the galley where one of the Grey Hound's crew asked in a general way – Have you heard the news? – To which I replied No, and thereupon he said in the presence of myself and* **Chas. Story** *and* **Will. Naylor**, *two of our crew, and several others to the number of four or five (whose names are* **Geo Short**, **James Town**, **Robert Ticknor** & **Joseph Bennett**) *that the Grey Hound was off Poole Bar lying too, about a night or two before, as I understood him (though I do not recall any precise day) when a boat came upon the quarter and asked the man that was heaving the lead to give them a rope which the man leaving off heaving the lead, did throw them a rope, but what kind of rope he did not say. That the vessel having a great motion broke one of the sweeps which projected over the stern, against the boat. That* **Capt. Watson** *then came upon deck, and looking over the Taffrail* [The rail or upper part at the stern of a ship] *Damned his men's Eyes and ordered them to go forward and do their duty, for that it was only a Man of War's boat and let them go to Hell. That he ordered his foresail to be let down which being done, the vessel having fresh way through the water, the people in the boat either cast off the rope or that the fore thwart or whatever it was fastened to in the boat gave way, and set them adrift, whereupon one of the men in the boat cry out Lord Have Mercy Upon Us, and they never saw them afterwards*

2. *What reply did you give to this man after he had given this long account?*

   I made no reply at all.

3. *What reply did* **Chas. Story** *or* **Will. Naylor** *make?*

   I did not hear any.

4. *Were any questions asked by* ***Short, Town, Ticknor*** *or* ***Bennett****?*

   I did not hear any.

5. *Was any questions asked by any other of your crew that were present?*

   I did not hear any.

6. *Name the person who gave the account you have stated.*

   I cannot.

7. *Describe the person and say whether you should know him again amongst the crew if they were all produced.*

   I cannot, nor should I know the person if he was produced.

8. *What became of you when this conversation was over, and about what hour of the day or night was it when it passed?*

   It was about five o'clock in the afternoon to the best of my knowledge, and immediately after I went to bed leaving the Grey Hound's people in the galley, when the conversation passed.

9. *What impression did this conversation make upon you, did you consider it as a true or an idle report, and when and whom did you first relate what you had heard?*

   I was very much shocked, but I could not tell whether it was a true or idle report; however the next morning as soon as I saw Mr ***Butler****, I communicated to him, and as nearly to the best of my recollection in the very words I have now stated it.*

10. *What passed between you and Mr* ***Butler*** *upon you communicating this story to him?*

    Mr ***Butler*** *said it shocked him if so be it was true.*

11. *Did Mr **Butler** say that he had the same report from Mr **Carter**, the Mate of the Grey Hound in the cabin, or from any other person previous to the account you gave him of this accident?*

    No not to the best of my recollection.

12. *To whom did you next communicate this account?*

    To nobody, to the best of my knowledge,

13. *Did you never mention it to **Chas. Story, Will Naylor** or any other of your crew or had you no conversation with any of the crew after you got up?*

    No I did not, nor have any conversation with him or any of them.

14. *When was the first time after you had made this communication with Mr **Butler** that you had any further conversation with him upon the subject?*

    I never had any conversation with him afterwards, upon the subject.

15. *Do you consider yourself as capable from the voice, complexion, size, dress or any other particulars, to point out the person whom you have stated came out of the Grey Hound's boat on board your cutter, and gave the account stated by you now upon oath?*

    No I am not capable. I am certain I cannot.

16. *Was this man the only speaker upon the subject among the whole of the cutter's crew that came on board, or were there more who spoke upon it in your hearing, and if so how many of them spoke of it, and who are they?*

    There were three or four more of the crew who spoke to the same purport in my hearing at that time, but I made no reply to them nor do I know their names nor could I describe their persons, nor should I know them again if they were to be produced.

17. *You having sworn in your answer to question No 14 that you never had any communication with Mr* ***Butler*** *after the finishing time of your mentioning this subject to him, look upon this paper now delivered to you marked A, and say whether ever you saw the said paper before, or heard the contents ever read to you by any person?*

    No I never saw the paper before nor ever heard its contents read.

[The document then records an interjection by **Captain Watson** to Mr **Butler** and questions continuing the numerical sequence, each signed individually by **Butler**.

18. *How long have you known me?*

    I *think nearly four or five years.*

19. *Have you assigned any reason to any person for my sailing from Spithead at the time or about the 27th of January last, when it is alleged the Spitfire came thither?*

    [**Butler** responds to the inquiry panel members directly]:-*I cannot assign any reason for* ***Capt. Watson's*** *sailing from the Spithead at that time, but I thought it was very extraordinary that she should go away when I was informed the Spitfire came to an anchor just a head of the Hound; and I believe I may have mentioned this matter to some other persons but to whom I do not recollect*

20. [Cross questioning continues by the Surveyor's General:] - Do *you mean by the expression' very extraordinary' that you made in the preceding answer to impress the Surveyor's General with a belief that* ***Captain Watson'*** *so doing was a proof of a guilty conscience?*

    No Sir.

21. *Did you never express yourself to any person whatsoever that* ***Capt. Watson's*** *such sailing was a proof of a guilty conscience?*

    No Sir

22. *Did you not tell **Capt. Roberts** of Shoreham so?*
    No Sir

23. [Question by **Captain Watson** to **Butler**]:-*What opinion have you of my character and have you ever heard of any acts of inhumanity committed by me?*
    [Answering directly to the inquiry board] *No I know nothing prejudicial to **Captain Watson**'s character nor did I ever hear of any acts of inhumanity committed by him.*

24. [**Butler** questions **James Pain Bundell**]:- *Whether you do not recollect that after I received the callers order to come to London, my calling you down into the cabin and saying as I knew not my business in town, it struck me that it might be relative to what had been related respecting Capt. Cook and whether I did not read to you a paper marked A, now produced.*

    [**James Pain Bundell's** reply]:- *Yes you did and read the paper in the presence of **Naylor, Story** and myself more than once; and asked in particular whether all you had read was not true to which I replied it was.*

Taken & sworn this 7th March 1794 before us – Adjourned till Wednesday 12 Instant.

The last answer was signed by Bundell and the last page countersigned by **Burrow, Hammond** and **Barker**.

Transcript of entry for Wednesday 12 March 1794

*Wednesday 12th March 1794 – Present Mr **Burrow** Mr **Hammond***

*Cross question to the last witness **James Pain Bundel** by the Surveyors General*

25. *You having sworn in your answer to the question No. 14 that you never had any conversation with Mr* ***Butler*** *your Mate, upon the subject of Capt. Cook's death after you had made the first communication of what you had heard thereon to him. How come you to give in your answer No. 24 an account so contradictory to what you had sworn before?*

    I did not recollect the circumstances of Mr ***Butler*** *calling me into the cabin and reading over the paper marked A.*

26. *Are you sure that every answer you have given to the questions put by us and to which you have signed your name, are founded in truth and perfectly free from any mistake or want of correction as stated in exculpation of yourself in the last answer?*

    Yes and I abide thereby.

27. *Questions from the Surveyors General to* ***Charles Story*** *12th March 1794*

    Charles Story *a Mariner on board the Hound Cutter aged 24 years being produced and sworn as a witness to support the charge in presence of the accuser and accused answers upon oath the following interrogations.*

28. *Give an account of what passed aboard the Hound Cutter, on or about the 26th January last, betwixt you and any of the crew of the Grey Hound's boat that came on board your vessel, relative to the death of Capt. Cook and where was your vessel then at anchor & where was the Grey Hound?*

    Both vessels were at anchor at Spit Head. I do not recollect the precise day, but it was about the latter end of January, when the Grey Hound's boat came on board the Hound, where I then was; when one of the men that came in the boat, which was rowed alongside by six mariners, and a sitter (the sitter being Mr ***John Carter****, the mate of the Grey Hound) and I being upon deck when* *they so came on board, said Carter went down into the cabin to Mr* ***Butler****, our Mate, when one of the sailors who came in the boat, asked in general terms such of our crew as were then at hand which were myself,* ***William Naylor*** *&* ***James Pain Bundel****, and others whose names I do not recollect, 'have you heard the news?' To which replied 'What News?' And he answered concerning Capt. Cook and his boat crew being drowned; I then asked him how it happened and he said 'we were lying to off Poole Bar (Meaning the Grey Hound Cutter) when Captain Cook's boat came up under the said vessel's quarter, and some person in the said boat asked for a rope, and one of the people in the cutter threw him a rope, & there being so much sea, and the cutter having so much motion, they supposed one of the sweeps struck the boat and was broken, whereupon they called another rope, and the man at the lead ran aft to get another, when* ***Captain Watson****, coming upon deck, Damned him for a Son of a Bitch & bid him go forward about his business; and looking over the cutter's stern, said 'It is only a Man of War's boat & let them go to Hell; and ordered the man to let draw the foresail by which the cutter gaining fresh way drew the boat so violently through the water, that the either the thwart broke, or they cast off the rope for fear of being drowned under the water; and the people on board the cutter heard the people in the boat say 'Lord have mercy upon us' & they never saw them more.*

29. *What reply did you give to the person giving the account you have just stated?*

    I said it was a shocking thing if it was true.

30. *Did you ask him any other questions or did any of your crew then present ask any such?*

    I did not, nor can I tell whether any other questions were asked by any persons present, not having heard any such.

31. *Name the man that gave you this account.*

    I cannot.

32. *He is a young fellow about my height five foot six inches, pock marked, long black hair, had on a canvas jacket, rather dark complexion, his voice I cannot describe nor his eyes, it being in the evening, his hair being then tied, but upon the whole I think I should know him again if he was produced.*

33. *Did he say whether Capt. Cook when he came alongside in the boat asked to come aboard the Grey Hound?*

    I cannot justly say whether he did or not.

34. *Did you communicate what you had heard immediately upon the Grey Hound's boat leaving your cutter to the Mate of the Hound, or to any or what other persons?*

    I did not to the mate but we might talk it over among ourselves.

35. *When was the first communication you made of it to the Mate; and was it the same night that you heard it?*

    It was not the same night that I heard it, nor can I justly say when it was that I first told him of it.

36. *Was the Mate in the cabin, in the hold, or upon deck when you first told him of it?*

    I cannot justly say whether it was in the hold, or upon deck, but it was not in the cabin when I first told him.

37. *Who was present when you first told him?*

    I cannot say particularly, it might be in the presence of most of the ship's company.

38. *What did the Mate say when you first told him?*

    I cannot tell whether he made any reply or not.

39. *You having said in your answer No 29 that it was a shocking thing if the account was true; why did you take upon you to conceal this thing from the mate, and not instead inform him of it the moment that you heard it, or at least in the course of that night.*

    I have nothing to say to it.

*Cross questions to Mr **Butler** by the Surveyors General*

40. *Wither do you recollect when and upon what day did the preceding witness give you any account of what he had relative to Captain Cook's death.*

    *It was not the same night, neither can I say what particular day nor who was present at the time; but it was previous to my examining them upon the paper I have produced marked A and previous also to the account I gave to **Captain Roberts.***

*Question by the Surveyor General to **Storey**.*

41. *Did Mr **Butler** ever call you down into the cabin of the Hound Cutter and read a paper to you which is now shown to you, and in whose presence was it read to you?*

    *I believe the contents of the paper now shown me marked A were read by Mr **Butler** in the cabin of the Hound in the presence of James **Bundel,** it being twice read over to me and the said **Bundel** and we agreed with what we had heard.*

42. *Was **William Naylor** present at this reading of the paper; and how long since may it be that this paper was read to you?*

    *No he was not nor any other person excepting Mr **Butler, Blundel** and myself, and it was about a fortnight ago.*

43. *What conversation passed betwixt you & Mr **Butler**, when he brought this paper into the cabin of the Hound & sent for you by **James Pain Bundel** and you in consequence if such message attended him?*

    *When I came into the cabin Mr **Butler** said he had a paper to read to me, and he read it over twice and asked me if it was the same that I had heard & I told him it was, and thereupon he ordered **Naylor** to come to him.*

44. *As you have declared that you cannot say at what distance of time it was after you heard the account, that you communicated the same to Mr **Butler** can you assign any reason for you not doing so, by Mr **Butler**'s keeping you, or any of his crew at such a distance as that you could not have access to him at any time you wished for the purpose of communication?*

    *I might have gone to him at any time I chose, Mr **Butler**'s behaviour to the crew being very agreeable.*

Taken, sworn the 12th March 1794 before [signed by **Burrell** & **Hammond**] – *Adjourned till tomorrow morning at 10 o'clock.*

Transcript of entry for Thursday 13 March 1794

*Thursday 13th March 1794 – Present Mr **Burrow**, Mr **Hammond***

*Mr John **Carter** acting Chief Mate on board the Grey Hound but not sworn in as such; his commission now lying at Weymouth, and has been kept there for some time past: having been sworn, was then directed to withdraw in order that the examination of **William Naylor** on the part of the accused (who was then come) should be gone thro' which is the only evidence on his part to be produced. For his testimony Vide Page 53.*

***William Naylor** aged 35 years of age one of the mariners on board the Hound Cutter, being produced and sworn as a witness to support the charge in the presence of the accuser and accused sworn upon oath the following interrogations.*

45. Give an account of what passed on board the Hound Cutter on or about the 26th January last (betwixt you and any of the crew of the Grey Hound Cutter's boat that came on board your vessel, relating to the death of Capt. Cook; and where was your vessel then at anchor, & where was the Grey Hound?

***William Naylor** Examination. Both vessels were at anchor at Spit Head – When the Grey Hound's boat came on board the Hound Cutter, which might be between four & five o'clock in the evening some day in the latter end of January last (being rowed by six men and Mr **Carter** the Mate as sitter). The crew came out of the boat upon the deck, and I asked them if they liked to go below which some did and some did not. Three of them going below and three staying upon deck – myself also staying upon the deck when a person who came out of the said boat about my own height, having on a jacket, which might be brown or blue, I cannot say which, it being then dusk &whose name I do not know; he had long brown hair tied up in a ribbon, appeared to be a lightish complexion, with a shrill voice & whose person I should know again if he was produced; having also a pair of long flannel trousers; and this person after some short conversation approached me in the following words, or nearly so;*

He asked me first if I had heard the news. I told him no, asking him what news, and he answered concerning Capt. Cook and his crew and his death from being drowned. I asked him how it happened and he answered that they were lying too in the Grey Hound Cutter off Poole Bar. They discovered a boat full of people which came under the cutter's stern; and the man who had been heaving the lead, but whose name he did not mention, ran and gave them a rope, that there being a deal of sea under the cutter's stern by some means one of the sweeps was broken; and ***Captain Watson*** *coming up at the same time damned the man & told him to go forward and mind his business; and looking over the stern said it was nothing but a Man of War's boat & let them go to Hell; ordering the people at the same time to let draw the Topsail, which being done, the vessel having such way it was supposed they could not hold on in the boat; and a person in the boat was heard to cry 'Lord have mercy on us!, and were not seen afterwards – And the same person added that it was a serious concern and ought not to be concealed, therefore they meant to acquire two sorts of people, namely them that knew it, and them that knew it not – The said three people then went below, namely him that spoke & his two companions, into the galley; but I remained upon deck, it being my watch and continued there till eight o'clock before which time the Grey Hound people went away in their boat.*

46. *Had you any conversation with Mr* ***Carter*** *the Mate who was sitter in the said boat?*

    No I never spoke to him.

47. *What reply did you make to the person giving this account?*

    I answered him that it was a shocking thing if true, but I hoped it was not.

48. *To whom, and when, did you communicate this account, that you have related?*

    I might probably have mentioned it that evening to some of my ship mates but the next morning I told it to Mr ***Butler*** *the Mate, in the galley, it might then be about 7 o'clock.*

49. *What reply did Mr* ***Butler*** *make?*

    He asked me how I heard it and I told him by Capt. Watson's boat's crew and he said it was a very shocking thing if true.

50. *Did Mr **Butler** appear to be surprised at the account you gave him?*

    Yes Sir.

51. *Did he say he had heard it before?*

    No Sir.

52. *Did you understand from what passed then between Mr **Butler** & you that this was the first account he had heard of it?*

    I did so understand it, because it being the first of his getting up in the morning and I then going upon deck in my duty, had no further conversation.

53. *Give an account whether Mr **Butler** ever afterwards called upon and examined you relative to what you had heard, regarding this transaction; and when it was, and who was present at such examination.*

    *About the beginning of March to the best of my recollection, Mr **Butler** called me down into the cabin, and in the presence of **James Pain Bundel**, read a paper to me, and asked me if the content of it were what I had heard before; and I answered to him, to the best of my knowledge it was, nigh to what I had heard the boat's people say.*

54. *Look upon the paper now produced to you marked A & say whether it is the same paper that he read to you.*

    I think that it is.

55. *Have you had any conversation with **Capt. Watson** upon this subject since you came to London or with any of the Grey Hound's crew, or prevision thereto?*

    No, I have not.

Question to Mr ***Butler*** *by the Surveyor's General*

56. *Have you any further evidence to produce in support of this charge.*

    No I have not.

Question to ***Capt. Watson*** *by the Surveyors General.*

57. *Have you any questions to put, or cross examination to make of any of the witnesses whose testimony you have already heard taken?*

    No I have none.

Mr ***John Carter*** *called in and examined as follows.*

58. *Have you heard anything relative to Capt. Cook and his boat's crew being drowned, and if so give a full and explicit account of the matter?*

    I was aboard when the Spitfire's boat came along side of the Grey Hound, having been sent away upon duty in one of the boats together with six of the cutter's people; and did not return to the cutter till about six o'clock in the morning, having left the vessel the evening before, about nine o'clock – and upon my return I was told by ***Henry Buck*** *one of the sailors aboard the Grey Hound, and which was the first time I heard anything of the matter, that the spitfire's boat had been alongside the Grey Hound between 12 and on o'clock that morning; and he was fearful what was become of them because the weather was very bad. I asked him in what manner she came alongside and he told me that she rowed alongside the cutter & was alongside between the runner & tackle, and shrouds* [ropes or wires running from the top of the mast and attached to the side of the vessel[, *a considerable bit of time – That she had ropes hove into her from the cutter and that the boats painter* [a tow or tying rope] *was handed on board – That when the said boat* *came alongside they asked what cutter it was and some person replied the Grey Hound; and they in the Greyhound asked what the ship's boat was & they replied the Spitfire's boat; and they asked if the cutter was going towards Studland Bay, and they were answered from the cutter that they were going that way – That the boat by this time had veered away under the cutter's stern, then having the end of our main sheet on board, and there being a good deal of sea running, broke one of our sweeps by striking the boat – That she had nineteen or twenty fathoms of rope out – That the Captain gave orders to let draw and fill the vessel, and she having way thro' the water, they cast off the rope in the boat, and that* ***Charles Vie,*** *one of our people, saw them in the boat step the mast* [raise the mast] – *That on the 27th January last, the Grey Hound being at anchor at Spithead, and the captain on shore, I took the boat and came aboard the Hound Cutter that was at anchor about a quarter of a mile from us; taking along with me six of the sailors belonging to our vessel to row me on board, the names of whom were* ***Robert Vivian, Mark Loader, Samuel Marsh, James Harris, Edward Allen & William Bellringer****; in order to make enquiry if any smuggling was going on upon the coast of Sussex. That when I came there, I saw Mr* ***Butler*** *upon deck and he and I went down into the cabin and drank together a glass of Hollands* [Dutch Gin] *& water; and thereupon a conversation took place between us; first asking him what news, and whether he had heard of Cap. Cook and his boat's crew being drowned, to which he replied No; whereupon I told him nearly what I have before state, as told to me by* ***Buck****; and I further added in this conversation with* ***Butler*** *that* ***Buck*** *had told me, that he thought in his opinion that* ***Captain Watson*** *had not paid that attention to the boat as he had thought was requisite; and he was fearful what had become of her; I fancy I might say what had been told to him by* ***Charles Vie****, that he heard somebody in the boat say Lord Have Mercy Upon Us.*

59. *Did Vie say that he was upon deck when he heard that expression? Lord Have Mercy upon Us! And did he tell you whether the Captain was also upon the deck?*

    Yes, he did tell me that he was himself upon deck and that the Captain was also there.

60. *Did Buck say anything to you that when the Spitfire's boat was alongside that any of the people in the boat appeared to be intoxicated?*

    *I did not hear **Buck,** or any other person say they were intoxicated, nor have I heard since, any such account.*

Cross examined by Capt. Watson

61. *Did I or did I not tell you in the morning of the 25th of January, that the Spitfire's boat had been alongside our cutter and in the manner in which they acted?*

    *Yes you did after **Buck** had told me the story before related; you said that the boat had been alongside the cutter in much the same manner that **Buck** had related it.*

62. *Explain the words I said, in what I told you about the ropes, and the manner in which the crew of the boat conducted themselves.*

    You told me they had ropes from the vessel and you thought that some of them might have got on board the cutter if they wanted.

63. *Did I tell you that their boat when she came alongside did not ask for any assistance and therefore did not appear to be in distress not having asked for any?*

    You told me they did not ask for any assistance; but you did not tell me that you therefore thought the boat did not appear to be in any distress.

64. *Did I tell you she was a King's boat & I thought sent out upon a cruise after the smugglers, and would not have cast off if she had been in distress?*

    You said that when you found she was a King's boat you thought she was upon a cruise against smugglers,; but I do not recollect that you told me that she would not have cast off if she had been in distress.

65. *Do you not remember me saying to you that she never would have cast off if she had been in distress?*

    No, I do not remember it.

66. *Did you hear any of our crew say in the morning, before I came upon deck, that the boat was lost near the Grey Hound or in sight of any of her crew?*

    No, I did not.

67. *How long will a boat tow astern a cutter going close by the wind?*

    She will tow astern under a thrice or four reefed mainsail, but it depends upon the circumstance of wind and weather and also the cutter is making thro' the water, how long she will tow astern.

Taken and sworn the 13th March 1794.
[Signed by **Burrow** & **Hammond.**]

Adjourned till tomorrow morning at 10 o'clock.

Transcript of entry for Friday 14 March 1794

Present Mr ***Burrow****, Mr* ***Hammond*** *and Mr* ***Barker***

Mr ***John Carter*** *called in and again examined upon Oath Question by the Surveyors General.*

68. *Have you seen the Spitfire Sloop of War, and do you know any thing of the size of their boats?*

    Yes I have seen the Spitfire Sloop of War, and her boats row with six oars, and all six oar'd boats are nearly of a size.

69. *How long would such a six oared boat tow astern the Grey Hound Cutter?*

    It depends upon the circumstances and I refer to my former answer of yesterday No 67.

70. *Was the account you have related, committed to writing by you at any time after you had received it from* **Henry Buck**, *and have you communicated such account to any and what other persons since; besides Mr* **Butler**, *or do you speak from recollection only?*

    I speak from recollection only having made no memorandum in writing of the matter, and may have mentioned it as often as I have been asked to other persons besides Mr **Butler**; *and I particularly remember I mentioned it to Mr* ***Hansford*** [or **Horsford**] *a merchant and Mr* **Francis Stewart** *Gent' at which time there was a Lieut. of the Navy, who is upon the impress service at Weymouth, in company with them, many which Lieutenant said he had heard it before.*

71. *Do you come hither actuated by any degree of malice or with any view of promotion, or any other bias upon your mind against* ***Capt. Watson,*** *other than what is dictated by a pure and clear heart, wishing to declare nothing but the truth in the presence of God and the Surveyors General?*

    I come hither with no other view whatever but to declare the truth, the whole truth, and nothing but the truth, in the presence of God and the Surveyors General.

Cross questioned by ***Captain Watson***

72. *Whether you have had any, and what conversation; with any and which of the Grey Hound's crew, and have not heard some of*

them declare that they would say everything they could against ***Captain Watson*** *in order to procure his dismission; that they might be more at home?*

No, I have had no such conversation with any of the said crew; but I have heard ***George Edward Dodd****, the Captain's Steward say something to the same purport that is stated in your question.*

73. *What reply did you make to* ***Dodd*** *when you heard him make this declaration?*

    I made no reply, but just asked him this question – Did you hear any of the people say so - and he said he did!

74. *When was it this conversation happened?*

    I do not recollect the precise time, but I think it was since the order came for us to come to London.

75. *Did you mention this to the Captain or does the Capt. and you stand upon such a footing of distance as would not admit you to do so; or are you upon good terms; and have you received any obligations or benefit from the Captain or otherwise by his means, so that in point of friendship you might consider it as necessary to put him upon his guard against any malicious reports against him?*

    No I did not tell the Capt. what I recollected as he was not on board at that time; nor did I tell him at any time afterwards to the best of my recollection – Notwithstanding we are sometimes upon terms of friendship & at other times, the reverse; Altho' I must run myself under great obligations for his assistance in appointing me to the Mate of the Grey Hound.

76. *Altho' such obligations did not lead you to acquaint the Capt. with this matter did that silence proceed from your own inclinations, or from the advice of any other person or persons whatsoever?*

    From my own inclination, not having any advice from any other person or persons whatsoever, so to do.

77. *What grounds or reason were your inclinations to keep this a secret from the Captain founded upon?*

    I had no particular reason, and can say no more than that I did not acquaint him.

78. *Did not the Captain's Steward tell you the names of the persons who made the declaration above stated, and if he did not tell you their names, did you, or did you not ask their names?*

    He neither told me their names, nor did I ask him.

79. *Have you had any variance with Capt.* ***Watson*** *during the six years you have been with him, and if so state them?*

    We have had different disputes and variances, but the particulars I cannot recollect.

80. *Are you treated with civility by* ***Capt. Watson*** *does he admit you to the cabin to eat and sleep there, and does he furnish you with provisions, tea and sugar, liquors and at his expence?*

    I eat, drink and sleep in the cabin with ***Captain Watson****; and have been supplied with tea and sugar by him; & am entertained with better provisions than the men of the ship affords' and occasionally with liquors at his expence; but this is nothing more than what is common.*

The examination of ***Charles Vie*** *a mariner belonging to the Grey Hound Cutter, called by the surveyors General as to what he knows in support of the charges. Aged 37 years & has been in service 23 years at sea.* [**Vie** was evidently illiterate as each answer was signed only with his mark]

81. *Were you aboard the Grey Hound Cutter about the latter end of January last when a boat came alongside of the said vessel in which Captain Cook and some of his crew were?*

I was on board and was tending the main sheet as we were lying too, when the boat came alongside and the main sheet was thrown by one of the crew into the boat, by which she was veered astern, and I heard one of the people in the boat say 'what is the matter the man does not fill the vessel?' (In order to keep the cutter from dropping upon the boat) and ***Capt. Watson*** *being upon deck gave order to have aft the fore sheet which was immediately done; and then the boat towed, and just then they had either cast off the main sheet or else they had slipped it; they then stepped their mast and I heard one or more in the boat cry out 'Lord Have Mercy Upon Us'. I never saw them more; and after the boat was gone (about a quarter of an hour) the Captain gave orders to make ready to put the vessel about; and thereupon she missed stays* [To fail in going about from one tack to the other], *and after she fell off again, the Capt. ordered the sail to be hauled down to be reefed which made the fourth reef* [reef = rolling it down to reduce the area exposed to the wind] *in the mainsail & one in the foresail to be reefed, and after the sails were set, the vessel staid and she stood in the North Shore – That I recollect the morning after when Mr* ***Carter*** *the Mate was on board, I told him what I had heard called out from the boat of Lord Have Mercy Upon Us and this I said in the presence of the Captain. And I have had no conversation with any person but our own ship's company upon this subject.*

Cross questioned by ***Captain Watson***

82. Did you hear when you went aft to look in what situation the boat was, or at any previous thereto, any person in the boat call out for assistance, or desire to be towed to Studland Bay?

No I did not, but I heard ***William Buck*** *say the people want to know if* ***Captain Watson*** *is going towards Studland Bay, and I* *heard the Captain he should stand that way as soon as the vessel was under command.*

83. *What situation was the vessel in at this time?*

    She was under the chalk rocks and shingle, both being under our lee with a flood tide and a very heavy sea.

Witnesses called & produced on the part of the accused, ***Captain Watson****.*

The examination of ***Joseph Discott****, a mariner on board the Grey Hound Cutter taken on oath; called as a witness by* ***Capt. Watson*** *in order to prove the charge is false.*

84. *Give an account of what you know relative to Capt. Cook's boat and his crew coming alongside the Gray Hound Cutter some time in the latter part of January at which time you were at the helm.*

    I was at the helm about half after twelve o'clock at night when the cutter was standing to the west, with the wind about SW by W; and I heard a voice from windward, it blowing very fresh with a heavy sea, call out, but the words made use of I cannot say. I then looked and saw a boat bearing down right upon us with a lugsail up, this being about half past twelve o'clock at night, the Captain being then upon deck, who ordered the vessel to be hove about with her head off to pick the boat up, supposing her to be a tub boat and after the vessel was about, ***Capt. Watson*** *ordered me to put the helm down in order to lay the vessel too. – That after, as soon as she was laid too, for some minutes we lost the boat and soon after we saw her again, rowing towards us – That while I was lashing the helm down which I did by the Captain's orders, the boat had got under our lee quarter, and a man in the boat handed the bite* [loop] *of her painter*[rope], *and I and another man had it in our hands and we held fast until the main* *sheet was thrown into the boat. That while the boat was under the quarter and very soon after I had hold of the painter, I called out 'Shipmates why don't you come on board?' and a man in the boat cried out Lord Have Mercy Upon Us! That we let go the painter and the boat veered astern with the main sheet on board; the vessel still lying too. – The vessel coming in the wind, had stern way, and the ship drifted upon the boat, so that the said boat lay athwart the cutters rudder; and then the Capt. ordered me to put the helm up, in order to fill the vessel; & a Gentleman in the boat said 'The man will not fill his vessel! And* ***William Buck*** *made answer to the Gentleman saying 'The helm is up Sir' and after the vessel was full the Captain ordered the vessel to go forward and haul the foresheet aft in order to make full sail again, and after the foresail was let drawn, a Gent in the boat said 'Are you going towards Studland Bay?' at which* ***William Buck*** *said to* ***Capt. Watson*** *'The Gentleman is speaking in the boat & asking if you are going towards Studland Bay?' And the Captain replied to them 'I do not know but I may by and by, or words to that effect. That the cutter whilst she was underway and the boat in tow had three reefs in her main sail, a whole foresail and fifth jibs. - That the boat was a sea oared boat, cutter built, and that while she was coming alongside I heard some of our people say it was a Man'o War's boat but who it was that said so I cannot now recollect; and further, that I did not hear at any time while the boat was alongside or within hearing, any person therein express a desire to come on board the cutter – That the boat was gone about a quarter of an hour before the Captain ordered the vessel to be put about, and then she missed stays; and after she had done so he ordered the Mainsail to be hauled down and reefed, & also the foresail to be reefed, the fifth jib continuing – That before these orders were given, it came on a heavy squall and such weather that after the sails were set, we stood to the northward.*

85. *Did you hear the Captain say at the time of the boats departure or any time before that 'It is only a Man of War's Boat and let them go to Hell?' Or did you hear him say, such fellows were only a parcel of dammed Men of Wars rascals?'*

    No I never did hear him say so at that time or any other.

Cross Question by ***Captain Watson***

86. *What situation was the cutter in at that time?*

    In a dangerous situation, with the shingles & chalk rocks under our lee, and a flood tide.

87. *Did I not, or did you not hear me order ropes to be get ready, to heave unto the boat, before she came alongside?*

    You might say so, but I do not recollect it.

Taken & sworn the 14th day of March 1794 before us
[Signed] ***Burrow & Hammond*** – *Surveyors General,*

Transcript of entry for Saturday 15 March 1794

Present Mr ***Burrow****, Mr* ***Hammond***

Joseph Driscoll *being called in and for his examination answered the following interrogations put to him.*

88. *What age are you and how long have you been in the service, and how many years have you been a sailor?*

    I am about 28 years of age and have been in the service about 6 years in the Grey Hound and have been a sailor or seafaring man these 13 years.

89. *Name the man that threw the main sheet into the Man of War's cutter from the Grey Hound.*

    I cannot

90. *Was the Captain up on deck with you before you saw the boat, & at the time you heard the voice cry out, and after the boat was gone and how long did he continue so upon the deck thereafter?*

    He was upon deck at the several times mentioned in the question, and continued upon deck to the best of my recollection till half passed 3 o'clock that morning; being generally aft near me at the helm.

91. *When the boat was discovered, did the Captain call out to her?*

    No, not as I heard.

92. *Did he not call out to them in the boat to lower down their sail?*

    No, I never heard it.

NB. This person was picked out by **Charles Storey**, *from the crew of the Grey Hound's boat now passing in review before him, as the person who gave the account on board the Hound: which he has stated in Page 29, and the person alluded to in page 32*

The examination of **Samuel Marsh**, *22 years of age, and one of the mariners belonging to the Grey Hound Cutter; taken on oath, Has been about 11 years at sea.* [**Samuel Marsh** signed with his mark]

93. *Was you one of the crew that rowed Mr* **Carter,** *the mate of your cutter, on board the Hound Cutter, about the latter end of January end?*

    Yes I was.

94. *Did you go out of the board on board the Hound together with the rest of the boat's crew.*

    Yes we did.

95. *When you got on board, did you continue on deck or did you go down into any* and what part of the vessel?

    Yes we continued upon deck about a quarter of an hour (meaning myself and the boat's crew excepting Mr Carter the Mate, who went down into the cabin) and then we went down into the galley, which is in the fore part of the ship below, where the people dress their victual, eat & sleep.

96. *Had you had any conversation whilst you remained upon deck with any of the Hound's people?*

    No I do not recollect that I had any.

97. *Had any of your ship mates any?*

    Yes they had I believe.

98. *Had you any conversation with any of them in the galley, and with whom?*

    Yes I had but the person or persons that I conversed with, I do not know not being acquainted with any of the ship's company: but I believe there might be about 20 of us all together.

99. *Relate what you said among them in the galley.*

    After some previous conversation we began to talk about the loss of Captain Cook and his boat's crew; and I said that we saw the boat, and then put the cutter about, and she hauled her sail down and soon after came alongside, and a rope was hove into her from forward by ***Joseph Winter****, &* ***William Brett*** *also had another rope in his hand, but whether he threw it in or not, I do not know.*

100. *Describe the dress you were in the day you went aboard the Hound?*

I had a pair of canvas trousers on and a canvas jacket, and my own black hair tied & I had a round hat on.

101. *Look upon the witness* **Charles Story** *now produced and say whether you saw him on board the Hound Cutter and had any conversation with him.*

Yes, I did see him and had conversation with him.

Book B Being a continuation of the evidence taken by the Surveyors General upon the investigation of Captain Watson's Charge

Samuel Marsh *Examination continued*

102. *Relate the conversation faithfully that passed between you.*

I came in conversation with **Storey** *and I told him that I was afraid that Captain Cook and his boat's crew were lost, because she came alongside our cutter and that when we first saw her we were under way in Studland Bay, and we hove about for her, and we picked her up & there was a man standing by forward to heave a rope into her, but whether they got the rope or not, I could not tell; that she then dropped away under the quarter, and that the boat's crew then handed a bite of the boat's painter on board and the man who took hold of it desired them in the boat to clear it away and they said they could not – That I further said that while the boat was laying under the quarter that we were afraid she was stove with the crane over the quarter that she sweeps were put in, as one of the sweeps was broken. That I then said that she, meaning the boat, had dropped astern, we put our helm up after she had got the end of the main sheet and we let her draw the fore sail and a small time after the boat slipped. But I should have said* *before this that the Captain drove the people forward and said 'Damn your bodies go forward and haul the fore sheet aft, as it's a Man of Wars boat and never mind them' - That I further said that when the people had cast off the rope in the boat they cried out have mercy upon us.*

103. *Did you or did you not upon your oath say to* ***Chas. Storey*** *that* ***Captain Watson*** *said that it was only a Man of Wars boat and let them go to Hell.*

No Sir, I did not.

104. *Did you say that Captain Watson came upon deck and looking over the toprail damned his men's eyes & ordered them to go forward & do their duty for that it was only a Man of Wars boat and let them go to Hell?*

No Sir, I did not say so.

The examination of ***Henry Buck*** *a mariner on board the Grey Hound Cutter, taken on oath called for as witness over* ***Captain Watson's*** *charge but who could not have sooner on account of being in liquor & he being charged therewith certified he was ill.*

105. *Give an account of what you know relative to the unfortunate death of Captain Cook about the latter end of January last.*
That after 12 o'clock at night in the morning of the 25th of January we were cruising between Christ Church Head & the Needles, the needle light bore about SE&E distance about 4 miles with the wind about SW&S as nigh as I can now guess, we being then stretching about WNW or to the West of North, that some person forward in the vessel heard people call from a boat & I myself soon after heard them, being the walking abaft by the Companion [The companionway, A stairway on a ship leading from one deck to another], *we then made sight of the boat, but for my own part I did not see any sail upon her. That we then hove*

about, thinking that she was a tub boat & our head being then off we made her upon our lee beam. That we were then going to get out our boat, thinking she was a tub boat, but finding out she was a Man of War's boat by having a white bottom, and she coming forward, that business was dropped, and she rowed alongside of us to leeward, and struck us somewhere about the runner & tackle – That being alongside there were ropes hove in, but by whom I do not know, and after that they in the boat asked what cutter we were and having answered the Gray Hound some person in our vessel asked what boat that was, and answer was made the Spitfire (but whether the question' what cutter is that?' came from the boat before she came alongside or not, I do not recollect). That the main sheet was hove into the boat but by whom I do not know – They then asked us if we were going towards Studland Bay and answer was made to that we were going that way. We filled the vessel [To stabilise it by some means] *as soon as possible, the helm being up and it blowing very strong and we tried to stay the vessel and she missed. The vessel then had stern way and then I asked how does that boat come on astern, and* **Charles Vie** *made answer she is slipped and gone and I then said Lord have mercy upon them for I thought it was impossible for the boat to live.*

106. Did you hear any person call out in the boat 'why does the man not fill his vessel?'

No I did not myself hear any person call out but I heard my son [i.e. **Henry Buck's** son] *who was on board say so.*

107. How long do you think this boat lay under the dominion of your vessel?

She might be lying under the lee quarter about 4 or 5 minutes and when the vessel got head way she dropped astern and I never saw her afterwards and therefore cannot say how long she was there before she was adrift.

108. *You having said in your answer No.* [Number omitted] *that in reply to* ***Charles Vie*** *you said Lord have mercy upon them for you thought that it was impossible for the boat to live, such answer implies that you was very apprehensive for the safety of the boat & its crew, therefore I require you to declare upon your oath whether or no you think every thing was done that could have been done or ought to have been done by your vessel for the safety of that crew & its boat.*

    I think that if we had exerted ourselves we might have got some of the people up and saved them altho' I do not know that it was possible for us to save the whole of them.

109. *You having been a Chief Mate in this service on board the Revenue Cutter the Speedwell & the Wasp and also the Tamar commanded by* ***Captain Warren Lisle****, declare upon your oath, supposing you had commanded any of those vessels and had been in the precise situation and place when this boat came alongside of you, with the same wind and weather & with 28 men on board, which was the remainder of the complement then on board the Grey Hound, and under the same conviction you have expressed of the impossibility of the boat living, you would or would not have used any and what other endeavours & exertions to save the lives of these people, other than what was made use of on board the Grey Hound Cutter.*

    I should have done my endeavours to have saved them, meaning that I should have attempted to have got some of them out of the boat & particularly when the boat laid alongside which was the proper time to have got some of them out of the boat.

110. *Should you have been inclined to have done this as it was very bad weather whether or no the people in the boat had not called out for assistance?*

    I should have done my endeavours to save them & I do think I should whether they called for assistance or not.

*111. Did you or did you not hear **Captain Watson** call out to the people in the boat before she reached your vessel to lower down their sail?*

It might be said but I do not recollect it was.

*112. Did you or did you not hear **Captain Watson** make use of the expression or words to that effect 'they are only a parcel of damned Man of Wars Rascals & let them go to Hell?*

I never heard any such expression from the captain's mouth.

*113. Have you heard any of your crew say that such an expressions were in use by **Captain Watson** or any other words conveying any sense of contempt or conveying any meaning that he considered the boat in question & her crew as an object not worthy of his attention or even worth his while to give himself any trouble about them.*

*I cannot recollect the time but I do think I have heard such expressions passing amongst the crew but whether before or after we went to Weymouth I cannot say, but I heard **Captain Watson** say that if she was a Man of Wars boat she was out a cruising as we were, but I do not recollect that he ever made use of any contemptuous expression.*

*15 March 1794, Taken and Sworn before **Ed Burrow**, A Hammond.*

Adjourned till Tuesday Morning 10 o'clock for the purpose of arranging the evidence.

Transcript of entry for Tuesday 18 March 1794

*Tuesday March 18th 1794 – Present Mr **Burrow**, Mr **Hammond***

*Mr **John Butler** Mate of the Hound Cutter again called in and further examined upon oath.*

114. *You having in your examination on Friday the 7th instant, stated a transaction as happening on Sunday the 26th January last, but on being observed by* **Captain Watson** *that you was mistaken in that day, you replied you believed you was not, because you had taken it from your journal on board the Hound. However in the course of the investigation you did conceive yourself mistaken and it was altered from Sunday the 26th to Monday the 27th of January last, and the journal book being sent for & being now before us. Explain whether the transaction above alluded to is so stated in your journal as was alleged by you, & whether the Grey Hound's boat was on board you on Sunday the 26th or Monday the 27th of January last.*

Mr **John Butler** *further examination upon what he has sworn to in his former examination in Book A, Pages 1, 2,3,4,5,& 6.*

To the best of my recollection it was on Monday in the afternoon about 4 o'clock on the 27th of January last, the Grey Hound then lying near to us, as she had done the day before, on which day I had spoke to her as is stated in my journal; but it is not stated in my journal that the Grey Hound boat then came on board, because we never do make remarks in the journals when one boat passes from one cutter to the other, only by saying we spoke such a cutter, and having spoke the cutter on the 26 that had and so impressed my memory, that I thought it was on that day the boat came on board.

The examination of **Francis Bussell** *a mariner on board the Grey Hound Cutter taken on oath called as witness by* **Captain Watson** *in order to prove the charge as false*

115. *What age are you and how long have you been a mariner on board the Grey Hound Cutter & how long have you been at sea?*

I am 22 years of age and shall have been three years on board the Grey Hound next June and have been at sea 9 or 10 years.

116. *Give an account of what you know relative to Captain Cook's boat and his crew coming alongside of your vessel sometime in the latter of January Last.*

I went upon deck at 12 o'clock sometime about the latter end of January but I do not remember the precise day when I heard a noise & heard one of the people on board us say he saw a sail, and thereupon ***Captain Watson*** *ordered the boat's be got ready to be hoisted out, supposing as I imagine that the thought it was a smuggler, but perceiving the boat was rowing towards us we did not hoist out the boat; thereupon the boat came alongside and ropes were thrown on board her, and I helped to clear away one rope myself from forward to be thrown into her which I believe was thrown into the boat by* ***William Court*** *which rope was the bowsprit and shroud fall, about a nine inch rope which I think might be about 8 or 9 fathoms long. Hereupon I went aft and they in the boat handed a bite of the boat's painter board our vessel which was taken old* [sic] *of by* ***George Dawson*** *otherwise* ***Dodd*** *the Captain's Steward and I took hold of the bite from him. I did not see the man in the boat who handed the bite of the painter* [loop of the rope] *on board; at this time the boat was swinging right athwart the stern of the cutter, so that she must have been fast by some rope that had been thrown from the cutter before. I got hold of the painter but I do not know by what man or what ropes were thrown out to the boat betwixt the time that I cleared away the bowsprit shroud fall & the time that I laid hold of the boat's painter, it being dark & blowing weather so that I could not distinguish the people that throwed the ropes of the boats.*

117. *Did you set any and what time see this boat lay alongside the Greyhound and how long might she continue there?*

She came rowing stern on between the shrouds and runner & tackle and then swang alongside where she might lay the space of five or six minutes.

118. *During what space could any of the men have been got out of the boat on board the Grey Hound?*

Yes Sir, I think they might during the time she lay alongside on the quarter.

119. *Could any of the crew of that boat have got on board the Grey Hound as she laid alongside if they had been so inclined?*

Yes, I think they might some of them.

120. *Now when the boat lay athwart the stern and you being abaft, did you hear any & what questions asked out of the boat to the Grey Hound's people?*

Yes they asked whether we were going towards Studland Bay (we being then three or four miles off the Needles lying too, under a three reefed mainsail & fifth jib & foresail to windward, and ***William Buck*** *in my hearing spoke to* ***Captain Watson*** *who was then upon deck and had been so from before the boat came alongside, telling what they said from the boat which the Captain said 'I think I shall stand that way by and by' and this was spoke by the Captain to* ***William Buck*** *who passed the word to them in the boat in so loud a manner as they in the boat must I think have heard them.*

121. *Did you or did you not hear any such expressions come out of the boat which was lying athwart the stern as those following? 'Why does not the man fill the vessel?' or words to that effect, and also 'Lord have mercy upon us'?*

No, to the best of my knowledge I never heard any such words spoken.

122. *Did you or did you not hear any such expression as the following, made use of by any & what person on board the Grey Hound –*

'Lord have mercy upon them for I think it impossible for the boat to live'?

After the boat was gone I cried out myself and I also heard some others of the ship's crew cry out nearly the same words.

123. *Did you or did you not hear the following expressions made use of by* **Captain Watson**, *'They are nothing but a parcel of Man of War rascals and let them go to Hell' in words nearly to that effect?*

No I neither heard one or the other.

124. *You have said in your answer No 117 that the boat came rowing stern on to the Grey Hound when you first discovered her, what distance might she then be from the cutter when you first discovered her?*

When I first saw her she might be about 100 yards.

125. *Can you assign any reason why she came rowing alongside and not under sail?*

I can assign no other reason but that it was more convenient.

Cross question by **Captain Watson**

126. *Did you not hear me order the people in the boat to lower her sails down?*

No I did not.

127. *Did you hear me order ropes to be got clear before she came alongside the cutter?*

I never heard the Captain give any such orders but when we swathe boat coming for us I was forward and we did prepare ropes.

128. *How long did the boat continue to be towed by the Grey Hound main sheet after her sails were filled?*

In about a quarter of an hour orders were given to put the vessel about & she missed stays, but whether the boat continued in tow by the said main sheet the full of this quarter of an hour or which part thereof I cannot say for till I was reefing the mainsail, I never heard she was gone.

Cross question by **Captain Watson**.

129. *When the boat was first seen by the Grey Hound & you supposed her a tub boat could you have boarded her with your boat had you got her out as was ordered?*

I should not have liked to have gone in her myself and several others of the crew said the same that night or since.

130. *What size was this boat you had on board?*

A four oared boat.

Charles Vie being again called in and asked the following question

131. *Under what sail was the cutter while she was under way & had the boat in tow?*

She was under three reefed mainsail a whole foresail & a fifth jib.

Cross examination by **Captain Watson** *–* **Henry Buck** *being again called and further examined.*

132. *Could you have boarded Captain Cook's boat when you supposed her a tub boat & was getting your boat out for the purpose?*

I should not have liked to have gone in our boat.

133. What size is the said boat you are speaking of?

She is a four oared boat called a Jolly Boat.

134. What reason do you assign for saying you should not have liked to have gone in the boat.

Because I thought there was too much sea.

Taken, Sworn this 18th March 1794

[Signed by **Burrow** & **Hammond**]

Adjourned till tomorrow at 10 o'clock.

Transcript of entry for Wednesday 19 March 1794

Mr ***Butler*** *being again called in and sworn and his examination taken upon the 7th Inst being read to him, wherein he stated the conversation between Mr* ***Carter****, Mate of the Grey Hound & him wherein he says ' nor does this deponent recollect that he at that time did ask the said* ***Carter*** *how or in what manner the said accident happened, nor had this deponent wither the curiosity of presence of mind to ask the said* ***Carter*** *who continued in the cabin, along with this deponent nearly two hours (after drinking some grog together) any further questions, nor did any other conversation whatever ever happen between them relative to the death of Captain Cook. And the said John* ***Carter****, Mate of the Grey Hound being also called & sworn and his answer to question no 58 being read to him. wherein he swears that after telling to Mr* ***Butler*** *the story which he had heard & which is therein recited relative to the death of Captain Cook, goes on upon his oath ' And I further added in this conversation with* ***Butler****, that* ***Buck*** *had told me that he thought in his opinion that* ***Captain Watson*** *had not paid that attention to the boat as he thought requisite and that he was fearful what was become of her'. And said Mr* ***Butler***

& Mr ***Carter*** *being confronted one with another with the purpose of reconciling if possible a matter so positively asserted by one, and denied by the other, and the said* ***Carter*** *abiding by the truth of his declaration to Mr* ***Butler****, and insisting that he did make the addition to the conversation as before recited., which was told him by* ***Buck****. He the said Mr* ***Butler*** *was called upon to answer why he did not state that matter upon examination upon oath and why he omitted so material a part of his story, implicating in a high degree the conduct of* ***Captain Watson****.*

Replies. 'I do not recollect that any such conversation passed, because if it had I think I should first have made an inquiry among our people whether any such conversation had passed between the Grey Hound's people and them, before any account was related to me by ***Bundell, Story & Naylor.***

[Signed by **Butler & Carter**]

William ***Naylor*** *being called in and sworn and his answer to question No 45 being read to him (wherein he states a conversation that passed betwixt him and a person that came out of the Grey Hound's boat when she came on board the Hound the latter end of January last whose name he did not then know but who he now apprehends & thinks was* ***Robert Vivian*** *by seeing and conversing with him in presence of the Surveyors General) in which answer he states ' That such person told him it was a serious concern and ought to be concealed therefore they meant to acquaint two sorts of people, namely 'them that knew it and them that know it not'. And being asked whether he abides then by; and if to the best of his knowledge he thinks that these words were spoken by* ***Robert Vivian*** *who is now procured before him, replies that he abides by what he has sworn to, and verily believes the said* ***Robert Vivian*** *to be the person*

[Signed by **William Naylor**]

Robert Vivian *being hereupon called in and sworn, declares that he is of the age of 27 next January 7 and served his time to the sea, that thrice or four days after Captain Cook's boat had been alongside the Grey Hound Cutter to which he belongs (which was in the month of January last, about the latter end with William* ***Bellringer****,* ***Mark Loader****,* ***Samuel Marsh****,* ***James Harris*** *&* ***Edward Allen*** *together with* ***John Carter*** [the] *sitter, went on board the Hound Cutter then at anchor at Spithead at about a quarter of a mile distance from the Grey Hound & found several of the Hound's people upon deck when they went on board and amongst the rest he entered into conversation with a man on the starboard side of the deck near the windlass to the following effect. That the Grey Hound Cutter was lying too off the Chalk rocks when a boat came alongside full of people just before or just abaft the runner & tackle, and that altho' I did not hear them call out for any assistance, I being then upon deck & near the point prepared with all the haste I possibly could to throw a rope to the people in the boat when* ***Captain Watson*** *called out to me '* <u>*It's a Man of Wars boat damn your body, what business have you with a rope – Away forward with you*</u>*', thereupon I made no reply but went forward to clear away the fore halliards, it blowing very strong, & expecting to hand in the sails for reefing – That we let down the foresail in order to fill her, and it might be then three or four minutes while the sails were drawing before she was thrown into stays and during which time she went monstrous fast through the water and I apprehend from knowing that one of our men had thrown the main sheet into the boat that the boat was in tow after the sails were full, that she might the whole, or in part time be so in tow, but how long a part of the time I can not ascertain, and this account I very might give to the witness* ***Naylor*** *now present and others on board the Hound Cutter, as I never made any secret of it till after I knew I was to go to London.* [**Robert Vivian** signed all answers with his mark]

135. *Do you recollect when you was on board the Hound Cutter saying that this matter was a serious concern & ought not to be concealed, therefore you meant to acquaint two sorts of people namely them that knew it, and them that knew it not?*

I cannot pretend to say whether I did or did not make use of such expressions but I have heard the words made use of but from or by whom I cannot say, nor can I positively say that I have not made use of them myself.

136. *Do you remember the following words being made use of either by any persons on board the boat or any person on board the vessel ' Lord have Mercy upon us' or 'Lord have Mercy upon them' or did you ever hear the Captain say 'They are a parcel of Man of War rascals' or 'Let them go to Hell'?*

No I never heard the Captain say those words, nor did I hear any such expression come from the boat, but after I found the boat was gone from one of our people, I myself cried out 'Lord have Mercy upon them' and being then talking among ourselves, others of our crew made use of the same expression.

137. *From the aforesaid expressions which you have recited being made use of, it seems as if you and the crew thought the boat & her crew was in imminent danger. Was that the fact?*

Yes we all thought that they never would see day light: I thought so, I know

138. *Under such an impression of imminent danger as you have described, declare upon your oath whether in your judgement these men and the boat could have been saved?*

Yes, some of them could if not all.

139. *Where was the boat when you was going to give her a rope?*

She was just under the counter.

140. *Do you know whether any other of the Grey Hound's crew had thrown a rope or ropes into the boat before you attempted to give a rope?*

Yes they had thrown the main sheet in & their painter was aboard our vessel.

141. *Who threw the main sheet in?*

I cannot pretend to say for I did not take notice of the person but I have heard it said that the main sheet was thrown into the boat.

Cross examined by Captain Watson

142. *When you expressed your fears that the boat could not see day light, did you not suppose as the wind was, and our situation then being so near the Chalk Rocks the boat would be lost there in attempting to pass them?*

It was from the heavy sea & the storminess of the night, together with the nearness of the Chalk rocks that my apprehensions that they would never see daylight arose and they being upon a lee shore. I thought they could never row the windward.

143. *Was not the cutter in great danger that night when the boat was alongside?*

I did not think the cutter was in much danger, but I thought the vessel was.

144. *Do you know where the cutter was & in what situation was she at the time the Spitfire's boat came on board?*

She was cruising between Christ Church Head & the Needles at the time the said boat came on board and then lying too. The Needles lights were under our lee but at what distance I cannot say.

S.G.

145. *Did you see the Needle Lights?*

Yes we did see them plain, the cutter head was to the southward & eastward with her starboard tacks on board.

S.G [Surveyors General]

146. *Which way did the tide seem at that time?*

It was a flood tide, and whilst I was helping to reef the foresail I wished we had done & the sails set because I thought we drove up to the north east too fast which might drive us askew or we might get into a heavier sea.

147. *From your answer to the preceding question it would seem as if you were apprehensive that you drove up to the north east too fast, by which the safety of the Grey Hound might be endangered. How long was it after the boat was gone that you made this observation.*

It might be better than a quarter of an hour after the boat was gone because I do not exactly know the time she went.

148. *As you have been asked about the situation the Grey Hound was in & the danger of her driving so far to the NE, declare upon your oath whether or no the safety of the boat & her crew was incompatible with the safety of the Grey Hound, or in other words would so much time have been taken up by the prevention of that boat & its crew as that the Grey Hound must thereby have been in the greatest danger, or that in case such crew had been saved the Grey Hound must inevitably have been lost?*

My answer is that the time it would have taken up to save the men would not have endangered the Grey Hound and that I always thought some or more of them might have been saved.

Mark Loader *one of the mariners on board the Grey Hound Cutter, being called in and sworn, answers upon oath to the following interrogations.*

149. Look upon the witnesses ***Will Naylor*** *and* ***Robert Vivian*** *and say whether you saw them in conversation aboard the Hound Cutter some time about the latter end of January last, and if so where about in the said cutter were they conversing?*

I did see them conversing together on the starboard side of the deck forward for some considerable time.

Sworn before us this 19 March 1794
[signed by **Burrow & Hammond**]
Adjourned till tomorrow at 10 o'clock

Transcript of entry for Thursday 20 March 1794

Thursday 20th March 1794 – Present Mr ***Burrow*** *& Mr* ***Hammond.***

Mr ***Henry Buck's*** *further examination.*

Captain Watson *being asked what persons on board his vessel could exactly ascertain her situation at the time Captain Cook's boat was first discovered, replies that Mr* ***Henry Buck*** *told him that he had set the Needle Light by the compass, thereupon Mr* ***Buck*** *is called in and sworn, and the following interrogations put to him, and his answers are as follows.*

150. Did you set the Needle Light by the compass at or before the time that the boat was first discovered?

I set the compass immediately upon coming on deck which was a little before we saw the boat.

151. *Describe the situation of the Grey Hound at the time the boat was first seen?*

    The situation of the vessel was then about 4 miles from the Needle Lights between the SE & E points and the ESE, the Shingles NE one and a half miles off and about one third of a mile from the Ridge of Christ Church Ledge, over which ridge (off which the vessel was then about) there is 6 fathom of water at low water and our vessel draws between 12 & 13 foot water only the depth of water decreases on both sides of the said ridge (viz) to three fathoms to the NW and to four fathoms of water to the SE.

152. *You having described the situation exactly of the vessel in your own preceding answer which situation is now declared by* ***Captain Bray*** *(present) to be nearly about it, and allowed by* ***Captain Watson*** *to be so nearly. Declare upon your oath whether or no the safety of the boat and her crew was incompatible with the safety of the Grey Hound or in other words would so much time have been taken up by the prevention of that boat & its crew as that the Grey Hound must thereby have been in the greatest danger, or that in case such crew had been saved the Grey Hound must inevitably have been lost?*

    We could have saved the crew and not have endangered the vessel.

Cross questioned by Captain Watson.

153. *Did the boat ask for any assistance from us.*

    No not in my hearing.

154. *Does not all ships, vessels or boats when in distress make such kind of signal, sign, token or by word in calling for assistance to give warning to such people as may come near them to put them on their guard & attention to enable them to give every kind of assistance in their power to those in distress?*

    Yes or ought so to do.

155. *How long have you been at sea in the service of the Revenue?*

I have been in the service about 17 years.

156. *I make no doubt you have had many boats alongside of your cutters in that length of time both by day & by night in strong winds and fine weather – Did you ever haul any of the people out of those boats, or had you a right so to do without their calling for assistance or wanting to come on board?*

I have had several boats alongside in different kinds of weather, but I do not recollect meeting with any thing of that kind in so much sea & blowing weather, and with respect to hauling the people in I never was in that situation nor did ever any thing of that nature come across me.

157. *Was not the vessel in a dangerous place at that time?*

Yes if anything had given way as having a strong flood tide.

158. *Pray where ought my greatest attention & duty to be at that time?*

Upon the Quarter deck on the weather side to give orders to your people being more convenient for the people to hear you.

S.G.

159. *Was the Captain in that situation?*

He was in that situation upon different tacks but I cannot take upon me to say that he was always so.

***John Carter** & **Henry Buck** confronted with each other*

***John Carter**, Mate, being again called in and his answer to question No 58 being read over to him in the presence of **Henry Buck** the person mentioned by him in the said answer wherein he states as follows - 'And I further added in the conversation I had with Mr **Butler** that* *the said* **Buck** *had told me that he thought in his opinion that* **Captain Watson** *had not paid that attention to the boat as he thought was requisite, and that he was fearful what was become of her' He abideth thereby & declares the said* **Buck** *told him so, who being confronted with the said* **Carter** *declares that he well recollects to have used the latter part of that expression Viz 'that he was fearful what was become of her' but that he cannot swear whether he did or did not make use of the former part of the expression not being able now to recollect it.*

[Signed by **John Carter** & **Henry Buck**]

John Harris *one of the mariners on board the Grey Hound cutter being called in and sworn answers upon oath to the following interrogations. First declaring that he is 27 years of age & has been for 8 or 9 years at sea, but was previous thereto a fishmonger.* [**John Harris** signed all answers with his mark]

160. *Give an account of what you know relative to Captain Cook's boat & crew that came alongside of the Grey Hound in the latter end of the month of January last & how you was employed at time and where the said vessel lay.*

161. *The vessel was about 4 miles from the Needle light when the boat was first seen & was sailing towards our cutter and I heard a noise calling out, but cannot tell what they said, and soon after the boat came alongside between the runner & tackle to leeward of us and just before the boat came alongside I was heaving the lead by the Captain's orders and had them hove in but once, at which last I found 12 or 13 fathoms water. Before the boat came alongside I saw two of our crew each of them throw a rope towards the boat, neither of which reached the boat, and seeing her drive astern, I threw down the lead to go to the assistance of the boat and called out to a man in the boat 'give me your painter into my hand'. I told him to clear it away & he said he could not, which painter not being long enough to make fast on board our*

vessel, two or three of our crew took hold of the painter and held fast till the main sheet was thrown into the boat and that when they hold of the painter I let go, being called upon by ***Captain Watson*** *who asked me where the lead was to which I replied, down there, to which* ***Captain Watson*** *answered 'Damn your blood what business have you there, the boat is nothing to you'. And thereupon I went and picked the lead up again to clear it away and before I had done so* ***Captain Watson*** *gave orders to let draw the fore sheet in order to fill the vessel and be ready about, which was done. The vessel in trying to heave about would not stay & just before that, the boat astern had slipped the main sheet, after having been in tow about five or six minutes, during which time the vessel had so much way through the water and drag'd the boat along with her so swiftly that they in the boat(as I imagine) were obliged to let slip the main sheet – After which the vessel missed stays and the Captain ordered another reef in the main sail and to have the foresail down & reef it.*

162. *Before you discovered this boat to be a Man of Wars boat there was an order given by the Captain for your boat to be got out, off the deck, but when it was discovered not to be a tub boat this order was not carried into execution, declare therefor upon your oath whether you should have been inclined to have gone off from your cutter into her in your boat?*

 No I should not, then being so much sea & blowing strong.

163. *Before the boat reached your vessel did you hear Captain Watson call out to them in the boat to lower their sail down?*

 Yes I did.

164. *Did you ever hear Captain Watson make use of the following expressions or words to that effect (viz) 'Damn them they are nothing but Man of Wars rascals let them go to Hell'*

 I did not.

165. *Could this boat & her crew have been saved by your vessel if proper measures had been taken for that purpose?*

 Yes they could.

166. *Had such measures been so taken would they have produced the loss of your vessel?*

 No I think not.

167. *Can you tell whether those in the boat let slip or let go of the main sheet.*

 No I cannot.

168. *Did the Captain order you to go from the lead?*

 No I went of my own accord to assist the boat.

169. *Did you hear any person in the boat call out for assistance?*

 No I did not.

170. *Did you hear the Captain order ropes to be got ready to throw on board before the boat came alongside?*

 Yes I did but we did not at that time know it was a Man of Wars boat nor did we know what boat it was.

Taken & sworn this 20th March 1794 before [signed by **Burrow & Hammond**)

Adjourned until tomorrow at 10 o'clock.

Transcript of entry for Friday 21 March 1794

Friday the 21st of March 1794

Present Mr ***Burrow****, Mr* ***Hammond****.*

The examination of ***George Edward Dodd****, Captain's Steward & mariner on board the Grey Hound Cutter – Near 27 years of age, has been at sea about 6 years & has been between 4 & 5 years on board the Grey Hound Cutter.*

171. *Give an account of what you know relative to Captain Cook's boat & crew that came alongside of the Grey Hound in the latter end of the month of January last, and how you was employed at that time.*

I think it was the 25th January about half past 12 o'clock in the morning about half an hour after the watch was relieved & I was gone down below to bed when I heard a noise upon deck. & hearing ***Captain Watson*** *call out from the deck 'lower down your sail, lower down your sail'. I put on my jacket & run upon deck, thinking some smuggling boat was at hand – That all hands were ordered to be called, but whether they were all upon deck before me I cannot say. The vessel then being between Christ Church & the Needles, lying too with her starboard tacks on board, the wind blowing from the SW. – That I believe I was the first person on board our vessel that saw the boat after the sails were hauled down, and then she was rowing up to our vessel, and I said upon deck I thought she was a Man of Wars boat as she rowed up so fast. Which I think was in the hearing of* ***John Harris, William Buck, Frances Bussell & Charles Vie,*** *as they were standing near me – That there were several people standing ready with ropes to throw into the boat which was done in consequence of orders given by* ***Captain Watson****, who previous to the boat getting alongside said in my hearing 'get ropes ready to throw into the boat' – The boat then rowed alongside and coming stern on struck our vessel between the after shrouds & runner & tackle. But whilst she was thus, whether any ropes were thrown into her I cannot say having thrown none myself but I do imagine that others did, though I* *cannot pretend to say that any such ropes were made fast on board her, or even reached the boat, for then she began to drop astern, but their boats painter was handed out & the bite of it was given into my hand by* **Frances Bussell***, and I imagine by the position the boat was in, that he was the first that got hold of it, and I handed it around the runner to* **John Harris** *as I believe, whereupon I heard* **Aaron Twistrap** *desire those in the boat to clear the painter away to which a reply was made that it was foul & they could not, and thereupon drifting astern the painter as I believe being still held on by our people, the main sheet was hove into the boat on which she dropped astern, & laying across the stern was under our sweeps, one of them was broken. – The vessel during this time having stern way till the helm was ordered to be put up by* **Captain Watson** *which stern way might near her boat, but I do not know that the cutter pressed upon the boat, altho' I was not at any great distance from the topsail nor will I swear that she did not.*

172. *How long upon your oath did the stern way that you have described continue?*

About four or five minutes.

173. *Did you not hear a voice from the boat say 'Why does not the man fill his vessel?'*

No Sir, but I have heard **William Buck** *say so since & from my own knowledge I heard Mr* **Buck** *reply from the deck to the boat 'Our helm is up' which was an answer proper enough to such a question as I know the helm was up. – After this I went more amidships in the vessel, but did not assist to let drawn the foresail, as I & some others of the starboard watch went below, at which time the vessel was making slow headway through the water and had done so about 10 minutes or a quarter of an hour, having the foresail to windward & during which time the boat was in tow.*

174. Look upon the paper now produced marked B and say whether you ever have seen it before, it being the Captain's answer to his charge?

Yes, I have seen this paper before it being all my own handwriting.

175. Did you dictate it as well as write it or did you only copy it?

I only copied it from a paper that was given to me by ***Captain Watson*** *which was not of his own hand writing, and I had no sort of concern in drawing it up, nor any further concern whatever with it excepting reading it to the people.*

176. Do you or do you not believe upon your oath that during the time that this boat was about your vessel that her crew might have been saved?

I think that most of them if not all might have got into the vessel if they had used any exertions.

177. Supposing these men in the boat incapable from cold, from liquor, or from fatigue to have used any exertions to have got into the Grey Hound, could you and the rest of your crew have by your exertions preserved these people's lives?

Yes, that most part of them I believe might have been saved by the exertions of our crew in hauling them into our boat.

178. Would the exertions that you mean in your last answer, have taken up so much time as would have endangered the Grey Hound, & whether or no the safety of the boat & her crew was incompatible with the safety of the Grey Hound, or in other words, in case such crew had been saved, must the Grey Hound inevitably have been lost?

No it would not have taken up so much time as would have endangered the Grey Hound if it had been done while the boat was as I have described alongside; but if it had not been done speedily as the vessel was then drifting to leeward of course the more time that was taken up in that service, the greater would have been the cutter's danger.

179. *You having said in your answer to question No 175 that you had no further concern with the paper then shown you, than reading it to the people, explain what you mean by the reading it to the people, what people do you mean. When was such reading & by whose orders was it done?*

It was read over by me to most of our crew in the cabin of the Grey Hound after she came up to Deptford by ***Captain Watson's*** *orders.*

180. *For what purpose was it so read?*

I suppose to let us know what he meant to say in his defence to a charge which I understood to be brought against him, and the Captain being present, said he desired nothing more from the people but to speak the truth.

181. *What reply did the people present make to the paper when they heard it read, did they say it was true or otherwise, or did they express any approbation or disapprobation of the contents of it.*

Some of them said when they came to be examined they would speak the truth, without observing whether the contents were true or false or expressing any approbation or disapprobation of the contents of it.

182. *Was that all that passed?*

Yes to the best of my knowledge.

183. *Give an account if you have heard any and which of your crew ever express any degree of malice or ill will towards* ***Captain Watson*** *or make use of any threats that they would say or do any thing in their power to get* ***Captain Watson*** *turned out that they might be more at home.*

Since the Grey Hound was ordered to Deptford I have heard ***William Harrington, Aaron Twiststrap*** *and* ***Joseph Harris*** *besides sundry others of the crew, whose names I do not recollect all presently make use of the following expression that they wished they knew anything that would turn him out & some of them whose names I do not recollect added that they might have more of the Sod.*

184. *What reply did you make to these people?*

 I answered that if he had offended upon any account on former occasions that it was a pity, that in case there was any grounds for former resentment, that such resentment should be exercised in their particular case, meaning by this that any inquiry that might be instituted relative to the death of Captain Cook, upon which subject they were then conversing and to this I do not recollect that they made any answer.

185. *Did you ever hear* ***Captain Watson*** *say they were only a parcel of damned Man of Wars rascals?*

 No I never did.

186. *Did you hear anybody in the Spitfire's boat desire that their boat might be born off and veered astern?*

 Yes I did hear one of the people in the boat say so but they did not ask for any assistance.

The examination of ***William Buck*** *a mariner on board the Grey Hound aged 22 years has been at sea about 8 or 9 years and has belonged to the Grey Hound 4 years next May. – taken on oath*

187. *Give an account of what you know relative to the Spitfire's boat coming alongside the Grey Hound the latter end of the month of January last.*

 A little after 12 o'clock at night I turned out to relieve the watch about half past 12 o'clock I being abaft heard hollowing &

looking over the starboard side I saw a boat under sail and I heard **Captain Watson** *call out to them to lower down their sail and gave orders to the cutter about which being done, I then saw the boat come rowing alongside, coming stern on, and struck our vessel somewhere about the runner & tackle, we being then lying too, and some of our people were preparing & did heave some ropes into her, but a bite of the ropes painter was handed out & taken hold of by some of our people. – That when the boat dropped under our stern, I cleared away & hove the main sheet in to her, I called out to the man in the bow of the boat to take a turn round their crosspiece to which he answered that it was gone. I then said make fast to your fore thwart which I suppose he did. The boat now being veered a little astern, some one in her asked 'What Cutter is that?' to which I replied 'The Grey Hound'. Then some none in the boat asked if we were going towards Studland Bay & thereupon I told* **Captain Watson** *what was asked and he replied ' We are going to reach that way'. And I passed his words to the boat, and further whilst the boat was lying under our quarter, the vessel had stern way & took the boat about the midship, and some person in the boat said' That man will not fill his vessel, & I replied 'Sir our helm is hard up', which it actually was. That soon after they in the boat had made the main sheet fast in her, the Captain gave orders to let drawn the foresail & the vessel then having swift way through the water, I looked over the stern & saw the boat in tow, which I observed for about five minutes. I then went forward to lend a hand to reef sails and the foresail drawing we had attempted to put the vessel in stays & that she missed stays and soon after that I heard the boat was gone.*

188. *Did you hear any body & what person say 'Lord Have Mercy upon them for what is to become of the boat' which expression implies she was in great danger?*

I think I heard **Charles Vie** *say so but I cannot be certain whether I did or not.*

189. *As you do not recollect whether any person said so or not what was your opinion, was the boat in danger or not?*

Yes I think she was in great danger but I did not say so to anybody.

190. *Under such apprehensions of danger as you have expressed, did the Captain make any attempt to go in pursuit of the boat or show any anxiety to recover her after she was gone?*

Not that I saw.

191. *Supposing these people in the boat had been in liquor, or fatigued with rowing or benumbed with cold, that they were incapable by their own exertions to get out of their boat, could you and the rest of your crew by their exertions have saved these people's lives?*

Yes I think they could.

192. *Would the making use of such exertions by your crew have been the cause of losing your cutter so that if you had attempted it your cutter must have been lost?*

No I think not, unless something had given way & then she might have been in danger.

Adjourned till tomorrow at 10 o'clock. – Taken and sworn this 21st March 1794 before us [Signed by **Burrow** and **Hammond**]

Transcript of entry for Saturday 22 March 1794

***Charles Vie. Samuel Marsh, Francis Bussel** + **Joseph Discott**'s further examination.*

*Saturday the 22 March 1794 Present Mr **Burrow** Mr **Hammond.***

***Charles Vie, Samuel Marsh, Francis Bussell** & **Joseph Discott** having been called in and sworn, and the following question having been put to them separately and apart, viz*

193. *Whether or no during the time that the Spitfire's boat was near to, or about the Grey Hound that the people in the said boat could or could not have been saved by your crew without endangering the said Grey Hound Cutter, to which they respectively answer.*

    Yes I do think so & without any danger to the Grey Hound.

[Only **Bussell** signed his name, all others made their mark in lieu of a signature.]

Questions to Charles Vie and Francis ***Bussel****.*

194. *Did* ***George Edw. Dodd*** *the Captain's steward when the boat was rowing towards the Grey Hound say to you that she was a Man of Wars boat?*

    No I never heard any such expression.

    [A margin note says see *page 49* a reference to previous evidence (question 171) by **Dodd**, namely '*…I said upon deck I thought she was a Man of Wars boat as she rowed up so fast. Which I think was in the hearing of* ***John Harris, William Buck, Frances Bussell*** *&* ***Charles Vie****, as they were standing near me'*. [The statement had later been underlined by the examiners, obviously after question 195 was answered]

Question to ***Joseph Harris*** *who is called in and sworn.* [all answers are signed by **Harris** with his signature]

195. *Did* ***George Edward Dodd*** *say to you when the boat was rowing up to the cutter that it was a Man of Wars boat?*

    I heard some body say so but cannot tell who it was.

196. *How old are you. How long have you been at sea & how long in the Grey Hound Cutter?*

    I am 24 years of age, have been 5 or 6 years at sea & 12 months in the Grey Hound Cutter.

197. *Whether or no during the time the Spitfire's boat was near to, or about the Grey Hound, the people in the said boat could not have been saved by your crew without endangering the said Grey Hound?*

    Yes I believe they might have been saved without endangering the Grey Hound unless she carried any thing away.

Question to ***William Buck.*** [Answers signed with his own signature]

198. *Did* **George Edward Dodd** *say to you when the boat was rowing up to the cutter that she was a Man of Wars boat?*

    I said that myself, but I do not recollect that ever he said so.

    [A margin note refers back to the answers given in question 171]

Cross questioned by ***Captain Watson****.*

199. *Did you hear anybody that was in the boat ask for assistance?*

    No Sir I did not.

200. *Did you hear any of them in the boat desire to be veered astern?*

    Yes I did hear someone in the boat say 'veer us astern, veer us astern' when she was laying just under the lee quarter.

201. *Do not you think the boat's crew could have got on board the vessel if they had chosen to?*

    I apprehend some of them might have got on board if they had chosen it.

SG [surveyors general questions continue]

202. *Under that apprehension I presume you mean to say that if the boat's crew were all of them not benumbed with cold, fatigued*

with rowing, or concerned in liquor, that some of them might have got on board if they had chosen it?

I mean to say that if they were all sober & not fatigued with rowing or benumbed with cold, some of them might have got on board.

203. *You having said in your question No 200 that you heard some one in the boat say 'Veer us astern, veer us astern' when she was lying just under the quarter, whether you apprehend that expression doubly repeated indicated any wish for the boat to be further removed so as to be out of the reach of the sweeps that projected over the quarter or from what other cause?*

I comprehend their so calling out twice they wished to get clear of the vessel to be out of the way of the sweeps.

***William Naylor** of the Hound Cutter being again called in and interrogated as follows:-*

A margin note indicates that this question is put by **Captain Watson** (even though it refers to the Captain by name as if asked by another). **Naylor** signed with his actual signature.

204. *When you was aboard the Hound Cutter & **Captain Watson's** boat's crew came on board as you have described in your examination wherein you have said that the man you conversed with said that they meant to inform them that knew it and them that did not know it; did that man or any other in your hearing express a wish that **Captain Watson** might be dismissed or turned out of the cutter, or in any other way express ill will or resentment to **Captain Watson**?*

Some person (but upon my oath I cannot say who) said it was a very serious concern, and that he ought to be turned out of the vessel, but whether such person was one of our crew or one of the Grey Hound's I cannot Say.

[Questions continue to be put by the Surveyors General]

205. *Did the man who was in conversation with you or did any other person belonging to the Grey Hound in conversation with yourself, or any other person in your hearing, say that they could have saved the crew of the Spitfire's boat that night that she was alongside the Grey Hound.*

No they did not say anything further than what I have stated relating to ***Captain Watson'*** *abusing them for going to the assistance of the boat.*

Charles Storey *&* ***James Pain Bundell*** *of the Hound Cutter being called in & sworn, the following question is put to them by* ***Captain Watson****.*

206. *Did you hear any of the Grey Hound's people that night ask when they were on board the Hound express any malicious intention, or declare that they intended to make use of the occasion to get* ***Captain Watson*** *turned out?*

So far they both answer No they did not. [Signed by both]

Question to ***Storey****.*

207. *Did you hear any of the crew of the Grey Hound say on board the Hound that night that they could have saved the crew of the Spitfire's boat that night that she came alongside?*

No I cannot recollect that any of them said so.

Question to ***Bundell***

208. *Did you hear any of them at that time say they could have saved the crew of the said boat that night.*

Yes I heard some person say that they thought they could have saved some of the crew of the Spitfire's boat but I do not know who said so.

Question to Mr ***Butler,*** *Mate of the Hound.*

209. *Did you hear Mr* ***Carter*** *when he was in conversation with you in the Hound Cabin express any malicious intent against Captain* ***Watson*** *or say that he deserved or that he* ***(Carter)*** *would use his best endeavours to turn him out or have you ever heard him say so since?*

     No I have not, neither then or since.

Robert Vivian *cross examination. –* ***Robert Vivian*** *being again called in and cross examined by* ***Captain Watson*** *upon oath the following interrogations were put to him.* [**Vivien** signed with his mark]

210. *Did you ever sail on board a Man of War.? How long did you do so & on board of what ship?*

     I did sail on board the Orestes Sloop commanded by ***Captain Chivers*** *& continued in her about nine months but never sailed in any other.*

[This appears to be HMS Orestes, an 18-gun Dutch-built brig-sloop of the Royal Navy, originally built as the privateer Mars. She was captured by the British in 1781 and later served in the Fourth Anglo-Dutch and the French Revolutionary Wars.]

211. *Was you ever employed to row in any of Man of Wars six oar'd cutters?*

     Yes I did belong to a six oar'd cutter.

212. *From your experience in that boat you have mentioned, declare upon your oath whether the Spitfire's boat, had the men been benumbed with cold, fatigued with labour, or concerned in liquor they could have been capable of rowing up to windward when the cutter was laying too in that sea?*

     If they had laboured under the incapacities before mentioned I do not apprehend they could.

213. From the time you first observed the boat to be rowing towards you, how long a time do you think elapsed before they fetched alongside the cutter?

I think it might have been 25 minutes.

Taken and sworn this 22nd March 1794 before [signed] ***Ed Burrows, A Hammond***

Adjourned till Wednesday the 26th instant to arrange the evidence and consider the same.

Transcript of entry for Wednesday 26 March 1794

Wednesday the 26th March 1794 –
Present ***Mr Burrow, Mr Hammond***

The same witness continued

214. During the time the boat laid alongside or near the Grey Hound, did you ever hear ***Captain Watson*** *open his mouth or hail the boat in any manner whatsoever?*

I did not.

Cross questioned by ***Captain Watson***

215. Did you hear ***Captain Watson*** *order any ropes to be thrown into the boat?*

No I did not.

216. Do you not think that the boat's crew of the Spitfire would have got into the cutter of their own accord had an officer been in the boat?

If I had been on the boat whether there would have been an officer or not in her, I would have got out, but I cannot tell what others would have done.

The examination of ***William Brett*** *acting as Second Mate on board the Grey Hound Cutter taken on oath – Declareth that he is 21 years of age, has been a mariner 6 years and has belonged to the Grey Hound Cutter about 13 months.*[**Brett** signed with a written signature]

217. *Was you on board the Grey Hound Cutter at the time the Spitfire's boat came alongside, & if so were you upon deck?*

     Yes I was and upon deck, which deck I did not leave for the space of one quarter of an hour after we saw the boat & she was towing astern when I left the deck rather slowly than rapidly.

218. *Whether or no during this time the Spitfire's boat was near to, or about the Grey Hound the people in the said boat could or could not have been saved by your crew without endangering the loss of the said Grey Hound.*

     Yes I believe a part if not all of them might have been saved without endangering the Grey Hound unless she had carried anything away.

219. *Can you by any observations or by any information you are possessed of, throw any light on the matter now under inquiry?*

     No I cannot.

220. *Did you hear the Captain order ropes to be thrown into the boat?*

     No I did not, I was at the other end of the ship.

221. *Did you hear any body in the boat ask to come on board or for any assistance?*

     No I did not hear any of them ask to come on board, but I heard one man in the boat cry out 'For God's sake give us a rope' and one was thrown.

222. *Do you know the man that called out & was it one of your own crew or was it one in the boat?*

 I cannot tell but I think it was one in the boat.

223. *Do you think that the boat's crew of the Spitfire would have got into the cutter of their own accord had an officer been in the boat?*

 Yes I think they would: at least I should have done so if there had been an officer in her.

224. *You having said in your answer No 218 that part if not all of the boat's crew might have been saved, by what means would you have saved them?*
 By hauling them out of the boat.

The examination of **Joseph Winter**, *a mariner on board the Grey Hound Cutter taken on oath –Declareth that he is in the 26th year of his age, has been at sea all his life from a child, and has belonged to the Grey Hound almost three years.* [**Winters** signed with a full signature]

225. *Was you on board the Grey Hound Cutter at the time the Spitfire's boat came alongside & if so were you upon deck.*

 Yes I was. [The answer is signed by **Joseph Winter, William Court, Aaron Twiststrap** and with the mark of **Joseph Dicker**.]

226. *Did you hear the Captain order any ropes to be thrown into the said boat?*

 No I did not, I was forward in our cutter and might continue upon deck about a quarter of an hour not belonging to the watch that was then set, having been called upon when all hands were called. [Signed with full signatures by **Joseph Winter** & **William Court**]

227. *Whether or no the Spitfire's boat was near to or about the Grey Hound, the people in the said boat could or could not have been saved by your crew without endangering the loss of the Grey Hound.*

I think that most of them could have been saved if not all, by hauling the boat up alongside again and getting the crew out. [This answer was signed by **Joseph Winter, William Court, and Aaron Twiststrap. Joseph Dicker** and probably **John White** made their marks. **William Harrington** also signed with his mark in the margin]

228. *Can you by any observations or by any information you are possessed of throw any light on the matter now under inquiry?*

No I cannot otherwise more than what I have said. [Signed by **Joseph Winter & William Court**]

229. *Did you hear any person in the boat call out for assistance?*

No I did not. [Answer signed by **Joseph Winter, William Court, Aaron Twistrap. Joseph Dicks** made his mark].

230. *Do you not think that the boat's crew from the Spitfire would have got into the cutter of their own accord had no officer been in the boat?*

Yes I think they would for their own safety. [Signed by **Joseph Winter & William Court**].

231. *When did you first discover the boat rowing & what time might she be coming alongside & what distance from the vessel?*

It was so dark that we could not see her above 200 yards from the vessel & she might be about 7 minutes rowing that space. [Signed by **Joseph Winters** and **William Court** and signed with his mark by **Joseph Dicks**.]

The examination of ***William Court*** *a mariner on board the Grey Hound taken on oath – Declareth that he is 22 years of age, has been at sea seven years & four years thereof in the Grey Hound Cutter.*

232. *Q & A The witness having heard the aforegoing questions put to* ***Joseph Winter*** *read and having given answers precisely agreeably thereto has signed the said answers in testimony thereof.* [Signed by **William Twistrap**]

Aaron Twistrapp's [sic] *Examination.*

The examination of Aaron Twistrap a mariner on board the Grey Hound taken on oath – Declareth that he is 25 years of age, has been at sea 10 years & has belonged to the Grey Hound 3 years & 8 months.

233. *This witness answers to the question No 225 having heard it read* <u>*in the*</u> *affirmation & also to the questions no 226 answers with this exception that he was abaft, instead of being forward, also to the question No 227* <u>*in this*</u> *affirmation.* [signed by **Aaron Twistrap**]

234. *To question No228 he replies as follows*

    I think that if we had not let drawn the foresail that the boat might have been hauled alongside so that the men might have safely got out or that we might have hauled them out.

235. *To No 229 he answers in the affirmative.*

    To the question No 230 he answers that whether there was an or was not an officer in the boat his opinion is that when the boat came first stern on upon the cutter that three or four of the men forward in the boat might have got out, but that after she veered astern they could not get out unless the boat had been hauled up again, which I think might have been done if we had not let drawn the foresail, and if the foresail had been kept aweather of helm a midship, the cutter would have not had so much way through the water & thereby the boat might have been hauled up & the people been saved.

236. *Q & A To question No 231 he answers that it was not a minute before she was alongside from his first seeing her as he was but just come upon deck.*

237. *Did you hear* **Captain Watson** *open his mouth & say any one thing to the crew of the boat while she lay alongside?*

    No I did not.

238. *Did you hear* **Captain Watson** *make use of any contemptuous expression such as '<u>they are only Man of War Rascals let them go to Hell</u>'.*

    No I did not.

The examination of ***Jos. Dicker*** *a mariner belonging to the Grey Hound Cutter, taken on oath – Declareth that he is 23 years of age, has been at sea betwixt 6 & 7 years & has belonged to the Grey Hound Cutter 15 months.* [**Dicker** signed all answers with his mark]

239. *Q & A Having heard the evidence of* ***Jos. Winter, William Court & Aaron Twistrap*** *read he agrees with the testimony they have given except to No 226 he declares he did hear* ***Captain Watson*** *give orders to get ropes ready before the boat came alongside to throw into her – and with respect to No 235 he perfectly agrees with the answer given by* ***Twistrap*** *– And also agrees with the answers to question No 234.*

240 *Q & A In answer to question No 238 he answers that he never heard any such expressions.*

The examination of ***Thomas Bagg*** *aged between 19 & 20 years taken on oath – Declareth that he has been about 12 months at sea & has belonged to the Grey Hound Cutter about 6 months.*

241. *Was you on board the Grey Hound Cutter at the time the Spitfire's boat came alongside the latter end of January last, was you upon deck & did you hear the Captain order any ropes to be thrown into the said boat?*

I was upon deck & did hear the Captain order ropes to be got ready before she came alongside but whether the said ropes were or were not thrown into her I cannot say because I did not throw them myself but I believe they were thrown in. [Signed with their marks by **John White, Thomas Bagg, William Barrett and William Harrington**.]

242. *Q & A To question No 227 which is now read to him & he understands the import of it – Replies, I think they might have been saved & the Grey Hound not been lost.* [Signed with their marks by **Thomas Bag & William Barrett**.]

243. *Did you hear the captain make use of any contemptuous expression of them in the boat such as 'Damn them, they are only a parcel of Man of War rascals let them go to Hell or did you hear him call for any assistance from the boat?*

No I did not. [Signed with the marks of **Thomas Bag, John White and William Barrett**.]

Wm. Barrett. *The examination of* ***William Barrett*** *aged 19 years taken on oath – Declareth that he has been a mariner 5 years and has been about 3 month belonging to the Grey Hound Cutter.*

244. *Q & A Agrees in answers given by the last witness in every respect to the questions answered as stated in No 241, 242 & 243.*

John White *Examination.*

The examination of ***John White*** *a mariner on board the Grey Hound Cutter taken on oath – Declareth that he is 34 years of age, has been* *a mariner 16 years & has belonged to the Grey Hound since January last.* [**John White** signed with his mark]

245. *Q & A The question No 241 being read to him agrees therewith excepting he did not hear the Captain give orders for ropes to be thrown into the boat.*

246. *Q & A To question 227 he replies that he thinks the boat's crew might have been saved without endangering the cutter.*

247. *Q & A In reply to question No 243 he says he never heard the Captain make use of any such expression.*

248. *Did you hear the people in the boat call out for any assistance? I did not hear them.*

William Harrington*'s Examination.*

The examination of ***William Harrington*** *a mariner belonging to the Grey Hound Cutter taken on oath – Declareth that he is in 34 years of age, has been bred up to the sea & has been near 3 years in the cutter Grey Hound.* [**Harrington** signed all answers with his mark]

249. *Q & A To question No 241 which had been read to him & he fully understands the import thereof, he replies that he was on board the Grey Hound & upon her deck at the time the Spitfire's boat came alongside, but he did not hear the Captain order any ropes to be thrown into her*

250. *Did you hear any body from the boat call out for any assistance or did you hear* ***Captain Watson*** *make use of any contemptuous expressions of those in the boat such as 'Damn them they are only Man of Wars Rascals let them go to Hell?*

     No I did not,

251. *Q & A To question No 227 which has been read to him & he fully understands the import thereof he declares upon his oath in his opinion that the crew might have been saved without endangering the Grey Hound.*

The examination of **Robert Swatridge** *a mariner belonging to the Grey Hound Cutter – Declareth that he is 22 years of age, has been 8 years at sea & 4 years on board the above cutter. Taken on oath.* [**Swatridge** signed his answers with a signature]

260. *Was you on board the Grey Hound & upon the deck when the Spitfire's boat came alongside your vessel some time in the month of January last – Did you hear them call out for assistance – Did you hear* **Captain Watson** *order ropes to be thrown on board and did you hear the said Captain make use of any contemptuous expression in respect to that boat such as 'Damn them they are only a parcel of Man of Wars rascals & let them go to Hell'?*

    Answers – I was on board & upon deck but I did not hear them call out for assistance nor did I hear Captain order ropes to be thrown, nor did I hear him make use of contemptuous expressions against any of the persons in the said boat.

261. *Declare upon your oath whether or no during the time* ***Captain Cook*** *& his boat's crew lay near to or about the Grey Hound, the people in the said boat could or could not have been saved by your crew without endangering the loss of the Grey Hound. Yes I declare I think they might without endangering the loss of the Grey Hound.*

262. *Did you hear Captain Watson say to the boat that it was the Grey Hound Cutter?*

    No I did not but I heard **William Buck** *say so.*

Taken, sworn the 26th March before [signed by Burrow **& Hammond**]

Adjourned until Thursday Morn. 27th

Book C Being a continuation of the evidence taken by the Surveyors General upon the investigation of Captain Watson's Charge

Transcript of entry for Thursday 27 March 1794

Thursday the 27th March 1794 –
Present ***Mr Burrow, Mr Hammond.***

The examination of **David Hibbs** *a mariner on board the Grey Hound Cutter taken on oath – Declareth that he was on board the said Grey Hound at the time the Spitfire's boat was alongside but that he never saw the said boat being below deck all the time she was there or know anything about her.* [**David Hibbs** signed with his mark].

The examination of **Mark Loader** *a mariner on board the Grey Hound Cutter taken on oath – Declareth that he is 21 years of age, has been 8 years at sea and has belonged to the Grey Hound Cutter one year & a half.* [**Mark Loader** signed with his mark]

263. *Was you on board the Grey Hound Cutter when* **Captain Cook***'s boat came alongside in the month of January last & was you upon deck?*

     Yes I was and upon deck.

264. *Did you hear* **Captain Watson** *order any ropes to be thrown into that boat.*

     I cannot pretend to say that I did or did not hear him, being forward upon the deck.

265. *Did you hear any of those in the boat call out for assistance?*

No Sir, I did not.

266. *Did you hear the* ***Captain (Watson)*** *make use of any contemptuous expressions of those in the boat such as ' Damn them they are only a parcel of Man of Wars rascals & let them go to Hell'?*

No Sir.

267. *Declare upon your oath whether or no during the time* ***Captain Cook*** *& his boat's crew lay near to, or about the Grey Hound, the people in the said boat could or could not have been saved by your crew without endangering the life of the Grey Hound.*

Yes, I do think as they might have been saved, without the loss of the Grey Hound.

268. *Explain your reasons for thinking so.*

My reasons are that if their boat had proved a smuggling tub boat, our Jolly Boat, the only one on board, must have been necessarily hoisted out, and that would have taken up more time than the saving this boat's crew would have taken up.

269. *Do you think* ***Captain Watson*** *would have ordered the Jolly Boat to have been hoisted out had it been a tub boat?*

I think he would.

270. *The supposed tub boat being upon your lee quarter could you have boarded her in your jolly boat?*

No we could not without running the risk of great danger.

271. *Supposing the Jolly Boat to have been hoisted should you have liked to have gone in her?*

No I should not, because there was too much sea but if the Captain had laid his positive commands upon me, I certainly should have gone.

272. *How long was the aforesaid boat rowing on board the cutter from the Grey Hound's heaving too?*

 About 10 minutes. [Answer was signed with the marks of both **Mark Loader** and **William Bellringer**]

273. *Did the boat crew of the Spitfire seem to row strong when they were rowing up to the Grey Hound?*

 I cannot say they did, because being dark I could not see. [This answer was signed with the marks of both **Mark Loader** and **William Bellringer**]

274. *Did the crew in the boat appear to be in liquor, or benumbed, or fatigued with labour & if so could they have rowed up to windward as they did when the Grey Hound was lying too?*

 No they could not. [This answer was signed with the marks of both **Mark Loader** and **William Bellringer**]

275. *Do you know whether the cutter was in a dangerous place or not?*

 If she stayed there all night drifting she might have been in danger but she would have weathered round the Needles with pleasure & have gone clear of all.

276. *Was the cutter lying too or sailing at the time?*

 She was lying too with her starboard tacks on board.

*The examination of **William Bellringer** a mariner on board the Grey Hound Cutter – Declareth that he is 24 years, has been a mariner going on 7 years & has belonged to the Grey Hound Cutter about 10 months.*

277. *Q & A The several questions No 263, 264, 265, 266, 267 being put to this witness and he fully comprehending their import answers to the same affect and purpose as the last witness & almost in the same words; but to question No 268 he answered*

as follows- My reasons are that after the boat was alongside, that if the vessel had been kept a little full & had no stern way, the men might with our assistance have been saved & this might have been done without endangering the loss of the vessel – To the questions 272, 273, & 274, he also agreed with the answers of the former witness – To question 275 he answers '- no, not just then I do not think she was in any danger.

278. *You having declared so positively as to the safety of the cutter. How did the Needle lights bear, at what distance and how, & in what direction of the compass did the chalk rocks seem from the above lights?*

 I can give no other answer but that I saw the lights under our lee bow.

The examination of ***Edward Allan*** *a mariner aboard the Grey Hound Cutter taken on oath - Declareth that he is 19 years, has been bred up as a fisherman, & has been in the cutter about 6 months.* [**Edward Allen** signed with his signature]

279. *Whether or no during the time the Spitfire's boat was near to or about the Grey Hound the people in the said boat could or could not have been saved by your crew without endangering the loss of the said Grey Hound*

 Yes I think they could, as we were lying too and had got ropes ready to heave into the boat which ropes was prepared by ***Captain Watson's*** *instructions, and when the said boat came rowing alongside our runner & tackle, I think we could then have saved them.*

280. *How long was the boat rowing alongside from the time you had first seen her rowing?*

 About a quarter of an hour.

281. *In what condition did you conceive the crew to be in, when the they came alongside?*

 They appeared to be tired.

282. *How do you know they were tired?*

 I supposed them so from the wind & them being wet by the foam spray of the sea.

283. *Did the boat's crew of the Spitfire appear to pull strong when they were rowing up to the Greyhound?*

 I suppose they were overjoyed when they saw the vessel & must have and must have pulled as strong as they could in order to get up to the vessel.

The examination on oath of ***John Boyes*** *a mariner belonging to the Grey Hound Cutter –Declareth that he is 59 years of age, was brought up a mariner & has been 5 years belonging to the Grey Hound Cutter.*
[**John Boyer** 'sic' signed with his full signature]

284. *Was you on board the Grey Hound Cutter when Captain Cook's boat came alongside in the month of January last & was you upon deck?*

 Yes I was and upon deck.

285. *Give us an account of what you know while that boat was alongside or near to the Grey Hound.*

 I first saw her coming rowing towards the Grey Hound & she was about ten minutes before she fetched alongside, which she did between the runners & tackle. I then saw some of our people preparing ropes to throw into her & I believe some were thrown into her & I saw her veering astern & she got under the counter and I heard one of the sweeps which was laying over the quarter crack, by which I apprehend it struck the boat – That afterwards she was in tow for about 10 minutes, and when we last drew the foresail, the vessel had great way by which she dragged the boat so violently through the water that they were obliged to slip the main sheet – I never heard any voice from the boat saying 'Lord have mercy upon us', nor did I ever see the boat any more.

286. *While the vessel was lying too and the boat came on board between the runner & tackle, could the people in her have got on board the cutter?*

 Yes, I suppose some of them might & if I had been there I should have got on board as soon as I could.

287. *Did you hear any person in the boat call out for assistance or express any apprehensions of danger?*

 No I did not.

288. *Whether or no the Spitfire's boat was near to, or about the Grey Hound, the people in the said boat could or could not have been saved by the assistance of your crew without endangering the safety of the Grey Hound?*

 Yes I think they might & without endangering the Grey Hound.

289. *Explain your reasons for thinking so.*

 The reasons I have for thinking the men might have been saved, are that if the people had been filled with her foresheet aweather, the boat might have been hauled up & the people got out, and without endangering the safety of the Grey Hound which I considered to be in no danger at all.

290. *Did the boat's crew of the Spitfire seem to pull strong when rowing up to the Grey Hound?*

 I suppose they pulled as strong as they could, but a man wet with the spray could not pull much.

291. *Did you hear anybody say from out of the boat that* ***Captain Cook*** *was in her?*

  No I did not.

292. *Two of your boats having been out that night & returned in safety between 6 and 8 o'clock in the morning, were any of them less in size than the Spitfire's boat?*

  They were less in size than Men of Wars six oar'd cutters & the Spitfire's boat that was alongside was a stouter boat than either of ours.

293. *As you have declared so positively that the vessel was in no danger at all, How did the Needle light bare, at what distance and in what direction of the compass did the Chalk Rocks run from the above lights?*

  I did not look at the compass and therefore cannot tell how the Need Lights bore and cannot say how the Chalk Rocks ran from the said lights.

294. *How were your two boats cruising?*

  I do not know where both of them were cruising but about 3 o'clock in the morning we fell in with one of them off Bagsham but we did not pick her up, because the wind had shifted more to the west off the land and the boat was then in more smooth waters by a great degree than the Man of Wars boat was when she lay alongside our cutter.

295. *Suppose the wind had not shifted and the boat had continued in or about the place where you saw her at 3 o'clock in the morning, was that boat in that situation (tho' not so stout a boat as the Spitfire's boat) in greater danger than the Spitfire's boat was when she laid alongside of your vessel?*

  I consider the Spitfire's boat to be in more danger.

The examination of **Stephen Watson** *a mariner on board the Grey Hound Cutter taken on oath – Declareth that he is 16 years of age, has been at sea between 3 or 4 years, and has been on board the Grey Hound about 2 years thereof.* [**Stephen Watson** signed with his signature].

296. *Give an account of what you know of what passed on board the Grey Hound Cutter on the night that the Spitfire's boat came alongside?*

 We were standing for the land in about half past 12 o'clock with our larboard tacks on board & we heard a voice hailing us, but I did not see the boat from whence that sound preceded and thereupon the Captain said he saw a boat under sail & immediately ordered the cutter to be put about and at the same time ordered ropes to be got ready to be hove into the boat, and then we saw the boat rowing upunder our lee & she came alongside near our runner & tackle which from the time we first saw her might be about 7 or 8 minutes and then I believe there were some ropes thrown into her & her painter was handed on board our cutter, then she dropped astern & as she dropped astern she struck one of our sweeps which was thereby broken, then the end of our main sheet was handed into her and they in the boat having made it fast, the boat dropped astern, and as I was standing aft I heard some person in the boat ask what cutter we were and they were informed by ***William Buck****, The Grey Hound. They then asked if we were going towards Studland Bay and the Captain answered that we were going to reach that way and soon after that I heard some person say we will let go by which I understand they were going to cast off the main sheet.*

297. *When the boat came first alongside the cutter could the people in her have got out of her into the cutter?*

 Some of them I think could but not all of them because the cutter boat had too much stern away.

298. *When you heard some person in the boat say we will let go as stated in one of your answers was the cutter then under way & the boat in tow & did she go so rapidly thro' the water as to reason any necessity of letting go.*

 Yes she was underway and the boat in tow but the vessel did not go very swift thro' the water because our foresheet was to windward.

299. *Did any of the people in the boat ask for assistance or ask to come on board the cutter?*

 No I did not hear any such request made.

300. *Have you by any observations or by any information you are possessed of thereon, throw any further light on the matter now under enquiry?*

 No Sir, no more than what I have said.

Taken and sworn this 29th March 1794 before us {signed by **Burrow & Hammond**.]

Adjourned till Tuesday next the 1st April 1794 for the purpose of.....

[The rest of the scribbled note at the foot of the page is faint and difficult to read, though it appears to says that it will re-examine the evidence of **Captain Watson** and others]

Transcript of entry for Tuesday 1 April 1794

Tuesday the 1st April 1794 – Present ***Mr Burrow & Mr Hammond****.*

Captain Watson *is called upon to state upon what foundation it is that the suggestion he has made to the Surveyors General that any combination or conspiracy against him by all or any of his crew is* *built for the purposes of getting him turned out of the cutter, that they might be more at home. And he alleging that had recovered information from* ***George Edward Dodd*** *his Steward & from* ***Thomas Bagg*** *one of the mariners belonging to the cutter to that effect, the said* ***Dodd*** *&* ***Bagg*** *are called in and re-examined upon oath.*

301. *give an account if you have heard any & which of your crew ever express any degree of malice or ill will towards* ***Captain Watson****, or make any threats that they would say or do anything in their power to get* ***Captain Watson*** *turned out that they might be more at home?*

Since the time the Grey Hound was ordered round to Deptford, I have heard ***William Harrington, Aaron Twistrap & Jos. Harris*** *besides sundry others of the crew whose names I do not recollect at present, make use of the following expression, that they wished they knew any thing that would turn him out and some of them whose names I do not recollect added that they might have more of the Sod.*

302. *What reply did you make to these people?*

I answered that if he had offended on any account upon former occasions that it was a pity that in case there were any grounds for former resentment that such resentment should be exercised in this particular case, meaning by this that any inquiry that might be instituted relative to the death of ***Captain Cook****, upon which subject they were then conversing, and to this I do not recollect that they made any answer.*

303. *Did you mention this matter to Mr* ***John Carter*** *the Mate, and when?*

I did mention it to Mr ***Carter*** *some little time after but whether it was upon the same day or not I do not recollect.*

304. *What answer did* ***Carter*** *make?*

To which Mr ***Carter*** *replied '****Captain Watson*** *deserves to be hanged & that I know it', alluding to* ***Captain Cook****'s death whereupon I said that he seemed to harken to any stories that were told against the Captain but would not listen to any that were in his favour and further I told him that what happened on board the vessel I had truly related to him of which he could not judge as he was not on board, thereupon he seemed angry and the discourse dropped.*

Thomas Baggs *was called upon as an evidence by* ***Captain Watson*** *but is not forthcoming owing (as is alleged by* ***Mr Carter*** *the Mate) to having met with an accident by which he is lame and unable to walk – he is therefore directed to be brought up in a boat tomorrow morning to which time this investigation is adjourned.*

Taken sworn this 1st day of April 1794 before us [signed by **Burrow & Hammond**] *Surveyors General.*

Transcript of entry for Wednesday 2 April 1794

Wednesday the 2nd April 1794 –
Present ***Mr Burrow & Mr Hammond.***

Thomas Baggs *being present and sworn in, interrogated as follows (viz)* [**Thomas Baggs** signed with his mark]

305. *Give an account if you have heard any and what which of your crew ever express any degree of malice or ill will towards* ***Captain Watson*** *or make use of any threats that they would say or do any thing in their power to get* ***Captain Watson*** *turned out that they might be more at home?*

I was walking the deck on board the cutter soon after she came into dock at Deptford & being upon the starboard side I observed

*Mr **Carter** our Mate walking upon the larboard side of the deck, along with one of our ship's company, and I heard the said **Carter** say to that person (whose name or person I cannot recall or know) 'I'm afraid the Old Man will sail with us again'; and the person answered ' Why' and the said **Carter** replied ' I do not think he will be turned out this time unless you all keep in one story'- Whereupon I gave no further heed to what they were talking about but went below, and I never had any conversation whatever with the said **Carter** or the person that was walking with him, or any of the crew relative to this matter till after I had told what I had heard to **Captain Watson**; which I did upon some day since the 7th of March last when I was in attendance with the boat's crew to be examined by the Surveyors General, but the precise day I can not ascertain. I recollect however that being left boat keeper (after bringing up the said crew at Kings Stairs, Tower Wharf, that **Captain Watson** came upon the wharf and called me to him out of the boat in the forenoon of that day, and thereupon I went out of the boat upon the wharf and the Captain asked me if I had heard **Carter** say that he wished the Captain might be turned out; I then repeated to the Captain what I have before stated.*

306. *Was there much wind that day when you heard **Carter** talking to the person you have mentioned, and from what quarter did it blow?*

     I cannot recollect.

307. *What distance were you from them when they were talking?*

     I might be 10 feet or more but I distinctly heard their conversation.

308. *Did Mr **Carter** ever ask you if you had communicated the conversation you had heard between him and the person to **Captain Watson**?*

     *One morning previous to our coming to town in the boat, being upon deck Mr **Carter** came up to me & asked me if I had told*

***Captain Watson** what I had heard him say and what it was that I did hear him say, for that he could not recollect, and I then told him that I had acquainted the Captain with what he had said & repeated to him the words I have before related and thereupon I walked away from him and he said 'I did not say so I believe, did I?*

Cross questioned by ***Captain Watson***

309. During the time the Spitfire's boat was alongside the Grey Hound can you tell how the Needle Lights bore & how the Chalk Rocks run from the Needle Lights by compass?

No I cannot.

S.G.

310. Q & A ***William Harrington*** *being called in and* ***George Edward Dodds*** *examination No 301 being read to him wherein he is charged with making use of the following expression that he wished he knew any thing that would turn* ***Captain Watson*** *out of the cutter, admits that he did say something to that effect and being asked his reasons for uttering such expression replies that he,* ***Captain Watson's*** *conduct that night in not saving the Spitfire's boat crew alongside the Grey Hound merited in his opinion his being turned out of the Grey Hound & it is his opinion still – That he believes this conversation happened at sea while the cutter was coming round to London, to the true of this he responded upon oath.* [**Harrington** signed with his mark]

311. Declare upon your oath whether you have ever heard Mr ***Carter*** *say to you or any others of the crew in your presence that they must all keep to one story?*

He never said so to me, nor to any other of the crew in my presence that I know of.

312. *Did you ever hear Mr Carter say to you or any of the crew 'I am afraid the old man will sail with us again'?*

No I never did.

313. *During the time the Spitfire's boat was alongside the Grey Hound can you tell how the Needles lights bore & how the Chalk rocks seen from the Needle Lights by compass?*

As to how the Needle lights bore I can not tell but the Chalks run from the Needles lights as I apprehend (the outer part of them) about NW but I never set them by compass.

314. *Did not the wind much increase soon after the boat was gone?*

I was not upon deck soon after the boat was gone, it being my watch below but I did not sleep, the vessel pitching about so much & the wind blowing strong and I recollect in about half an hour or an hour after the boat was gone, as I suppose it came on very squally & blew very hard.

Aaron Twistrapp's *re-examination*

315. *Q & A* ***Aaron Twistrap*** *being called in and* ***George Edward Dodd****'s examination No 301 being read to him wherein he is charged with making use of the following expression that he wished he knew any thing that would turn Captain Watson out of the cutter answers that those were not his words but they were as follows ' I said after having a general conversation passing amongst the crew (in which* ***George Edward Dodd*** *was present) relative to the death of* ***Captain Cook*** *& his people; that I wished it was in my power to turn him out, meaning* ***Captain Watson****, & this I said because I thought that* ***Captain Watson*** *did not pay that attention to the boat he ought while she was alongside & that he might have saved them if he would, & that was my opinion then & it is so now'. And to the truth of this he declares upon oath.* [**Twistrap** signed with his full name].

316. *Did you ever hear* ***Carter*** *say that he was afraid the Old Man, meaning* ***Captain Watson****, would sail with them again or did you ever hear him use the expression (viz) That if they did not keep all in one story he would not get turned out?*

No I never did.

317. *During the time the Spitfire's boat was alongside the Grey Hound can you tell how the Needle lights bore & how the Chalk rocks ran from the Needle lights by compass?*

No I cannot.

Joseph Harris

318 *Q&A* ***Joseph Harris*** *being called in and* ***George Edward Dodd's*** *examination No 301 being read to him, he denies that he ever made use of any such expressions or had ever any conversation with the said* ***George Edward Dodd*** *or any of the crew on board the Grey Hound as by him recited –but he recollects one day since the enquiry commenced being upon the stairs leading to this office that* ***Dodd*** *addressed* ***Twistrap*** *saying 'Lawyers, these Gentlemen (Meaning the Surveyors General) are a long time about it, I wish they would let you and I do it, we should soon have it done' to which I replied 'No that would not do for I suppose that one of you would be for keeping him in & the other for turning him out', meaning the Captain. And said* ***Dodd*** *&* ***Harris*** *being confronted said* ***Dodd*** *persists in his declaration & said* ***Harris*** *denies that he made use of such expressions as* ***Dodd*** *has accused him with.* [This is signed by both **Dodd** and **Harris**]

319. *During the time the Spitfire's boat was alongside the Grey Hound can you tell how the Needle lights bore & how the Chalk rocks ran from the Needle lights by compass?*

No I cannot.

John Carter *Confronted with* ***Thomas Baggs****.*

320. *Q & A Mr* ***John Carter*** *the Mate being called in and the evidence given by* ***Thomas Baggs*** *being read to him wherein he is charged with holding a conversation with one of the crew of the Grey Hound when she was in the dock at Deptford wherein the following expressions were used by him (viz) 'I am afraid the Old Man will sail again with us again' and the person answered 'Why?' and he* ***Carter*** *replied 'I do not think he will be turned out this time unless you all keep in one story' – Denies that he ever made use of any such expressions or ever expressed a wish that* ***Captain Watson*** *should be turned out, so far from it that he should be very sorry if he should be turned out, and this he now declares upon oath.* [**John Carter** signed with his full signature)

321. *Q & A* ***Thomas Baggs*** *being present and being confronted with Mr* ***Carter*** *declares upon oath that what he has stated with regards to the expressions made us of by Mr Carter is true & he abideth thereby.*

Taken before us this 2nd Day of April 1794 [Signed by **Burrow & Hammond**]

Adjourned till tomorrow at 10 o'clock.

Transcript of entry for Thursday 3 April 1794

Thursday the 3rd of April 1794 –
Present Mr ***Burrow*** *& Mr* ***Hammond****.*

John Carter *&* ***George Edward Dodd*** *Confronted.*

322. *Q & A Mr* ***John Carter*** *the Mate being called in, & the evidence given by* ***George Edward Dodd*** *in answer to question 304 being read to him whereon he is charged with making the following reply*

to said ***Dodd*** *to the said question No 304 '* ***Captain Watson*** *deserves to be hanged and that I know it' – and the further addition in the said* ***Dodd*** *evidence as follows being also read 'whereupon I said that he seemed to harken to any stories that were told against the Captain, but that he would not listen to any in his favour, & further 'I told him that what happened on board the vessel I had truly related to him of which he could not judge, as he was not on board, thereupon he seemed angry and the discourse dropped'. The said* ***Carter*** *being present and confronted with said* ***Dodd*** *declares upon his oath that he never made use of any such expression alleged by* ***Dodd*** *in his said answer, nor did* ***Dodd*** *ever make use of such expression to him (viz) that he seemed to harken to any stories that were told against the Captain but that he would not listen to any in his favour but he does admit that* ***Dodd*** *did say to him ' that you or no person could be a judge of the matter but those that were present'.*

323. *Q&A* ***George Edward Dodd*** *being present & confronted with aforesaid* ***Carter*** *declares upon his oath that what he has stated with regard to the expressions made use of by Mr* ***Carter*** *is true and also that his, the said* ***Dodd****'s reply thereto of his observations thereon are fully stated in his answer to the question No 304 and also true & he abideth thereby.*

324. *Did you after you had heard the expression aforesaid coming from Mr* ***Carter****'s mouth commit the same to* ***Captain Watson*** *or any other person or persons belonging to the Grey Hound?*

    I do not think I did.

Cross question to Mr ***Carter*** *by* ***Captain Watson****.*

325. *Have you never persuaded any of the crew to speak disrespectful things of* ***Captain Watson*** *or to keep on one story so that he might be got turned out?*

    No I never have.

20 of the Greyhound's Crew Examined as to the bearings of the Needle Lights & Chalk Rocks.

326. *Q & A The following persons being called in and questioned No 309 namely 'During the time the Spitfire's boat was alongside the Grey Hound can you tell how the Needle lights bore & how the Chalk rocks ran from the Needle lights by compass', respectively answer upon your oath? No.* [All men questioned signed their names in a list. Those that signed with a cross are indicated below with a cross between their forename and surname].

1. ***William Brett.*** *2.* ***Joseph Winter.*** *3.* ***William Court.*** *4.* ***Joseph X Dickens.*** *5.* ***Samuel X Marsh.*** *6.* ***William X Barnett.*** *7.* ***Edward Allen.*** *8.* ***John X Harris.*** *9.* ***Martin X Loader.*** *10.* ***William X Bellringer.*** *11.* ***Robert Swatridge.*** *12.* ***William Buck.*** *13.* ***John Boyer.*** *14.* ***Robert X Vivian. 15. Francis Bussell.*** *16.* ***David X Hibbs.*** *17.* ***Robert X Longman.*** *18.* ***John X White.*** *19.* ***James X Discott.*** *20.* ***Charles X Vie.***

327. *The several persons who have given answers to the last question are again called in and being required to declare upon their oaths respectively – Whether ever* **John Carter** *the Mate said to each or any of them the words which* **Thomas Baggs** *in answer No 305 charges the said* **John Carter** *with saying to one of the crew in his hearing as he was walking upon the deck "I'm afraid the Old Man will sail with us again' and the person answered 'Why?' and Carter replied 'I do not think he will be turned out this time unless you all keep in one story; Reply upon their oaths that* **Carter** *never made use of such expressions in their hearing or to any one of them individually.*

[The same list of mariners as listed in Q&A 326 are listed once more and signed in the same manner.]

20 of the Greyhounds Crew examined as to some disrespectful words spoken of the Captain.

328. *The several persons who have given answers to the last question are again called in and being required to declare upon their oaths respectively whether ever they heard any & which of the crew make use of the expression stated by* **George Edward Dodd**, *Captain's Steward, in his answer No 301 namely that they wished they knew any thing that would turn him (Captain Watson) out that they might have more of the Sod – Eleven of them whose names are here undersigned reply upon their oaths that they did not hear any such thing pass.*

[All men questioned signed their names in a list. Those that signed with a cross are indicated below with a cross between their forename and surname].

1. ***Mark Winter.*** *2.* ***William Court.*** *3.* ***Samuel X Marsh.*** *4.* ***William X Barratt.*** *5.* ***Edward Allen.*** *6.* ***John X Harris.*** *7.* ***Martin X Loader.*** *8.* ***Robert Swatridge.*** *9.* ***David X Hibbs.*** *10.* ***Joseph X Discott.*** *11.* ***Charles X Vie.***

And the following nine persons reply in the affirmation as follows (viz)

329. **William Brett** – *Yes, but I don't know who said so.*

330. **Joseph X Dicker** – *Yes I have heard such words but I don't know by whom spoken.*

331. **William X Bellringer** – *Yes but I do not know who spoke them.*

332. **William Buck** – *Yes – I have heard disrespectful words spoken of the Captain but I do not either recollect the words or know the persons that spoke them, but I cannot pretend to say on oath whether Mr* **Carter** *was or was not present.*

333. ***John Boyer*** *– Yes I have heard disrespectful words pass among the crew against the Captain in common conversation but I do not recollect the words or the persons that spoke them.*

334. ***Robert X Vivian*** *– Yes I have heard something like the words spoken but I do not know the persons.*

335. ***Francis Bussell*** *– Yes, it was the general talk of the crew that he deserved to be turned out for not taking care of the boat, meaning Captain Cook's boat.*

336. ***Robert X Longman*** *– Yes, that they would get the Captain out if it laid in their power but they gave no reasons and this he heard some of the people speak once or twice,*

337. ***John X White*** *– Yes that they wished the Captain was turned out and that something was said about the boat but he knows no particulars nor did they give any reasons.*

Taken Sworn this 3rd April 1794 before [signed by **Burrow & Hammond**] *Surveyors General.*

Adjourned till tomorrow morning at 10 o'clock.

Transcript of entry for Friday 4 April 1794

Friday the 4th April 1774 – Present Mr ***Burrow*** *& Mr* ***Hammond***

The following persons being called in and sworn being the two boats crew that were absent from the Grey Hound at the time the Spitfire's boat was alongside the Said Grey Hound and the following interrogations put to them (viz)

338. *Whether they ever heard any & which of the crew make use of the following expressions 'That they wished they knew any*

thing against ***Captain Watson*** *that would turn him out, that they might have more of the Sod or be more at home'.*

339. *Whether they ever heard* ***Carter*** *the Mate make use of the following words to any of them since the cutter has been in dock at Deptford – 'I am afraid the Old Man (meaning the Captain) will sail with us again', and the person whom* ***Carter*** *was then walking upon deck answering 'Why?' He replied to him 'I do not think he will be turned out this time unless you all keep in one story'.*

340. *Q & A* ***William Horn*** *a Deputy Mariner belonging to the Grey Hound & had charge of one of the boats that were absent in the night when the Spitfire's boat was alongside, being sworn replies that he never heard the expression mentioned in the first question made use of, but he has heard a general conversation regularly passing amongst some of them of which* ***Carter*** *was, that they wished the Captain might be turned out, but none of them assigned any reason to me except* ***Aaron Twistrap*** *who said that it was on account of the Captain's not paying such attention to the Spitfire's boat when she was alongside the cutter as he thought was necessary.*

[Signed by ***William Horn*** signed both answers with a full signature]

341. *Q & A To the 2nd question he replies negatively that he never heard any such words pass.*

342. *Q & A* ***John Miller*** *one of the boat's crew who was with* ***William Horn*** *on the night the Spitfire's boat was alongside the Grey Hound being sworn deposeth that he is 27 years last January & the same questions being put to him as were put to the last witness, he replies to the first of them that he never heard the expressions mentioned in that question made use of by any of the crew but he has heard the general conversation more than*

once, of several of them express a wish that **Captain Watson** *should be turned out & that he himself has said so, ' because that being in the Grey Hound's boat that same night off Great Haldiman and the cutter was standing right in, in much distress, baling out the boats with their hats & flashing their pistols with gunpowder as signals of distress, no notice was taken of them by Captain Watson's cutter, which came near to us within twenty yards although we called out loud enough for them to hear & they called back something which we did not understand and the cutter was then standing for the land & left the boat in the same circumstances of distress as when she came up to us.*

343. Q & A To the 2nd question he replies negatively that he never heard any such words pass.

[Both answers were signed with **John Miller's** mark, which like all the others was a simple cross.]

No 13
The testimony of this witness disclosing a new circumstance of which we never heard the least hint of before and implications **Captain Watson's** *conduct in a fresh matter which did not apply to the charge in question, altho' something similar to it. We forbear proceeding in the examination of any of the boat's crew that were along with* **Miller** *as to this matter as that would have tended to have fixed their evidence thereupon, when probably it may become the subject of a future inquiry.*

***Frances Tavills** aged 26 years a mariner on board the Grey Hound sworn.*

344. Did ever you hear **Mr Carter** *make use of the following expressions 'I am afraid the Old Man (meaning the Captain) will sail with us again', and the person with whom he was*

walking answering 'Why?' Said ***Carter*** *replied 'I do not think he will be turned out this time unless you all keep in one story'.*
I never heard any such words. [**Francis Tavills** signed with a full signature].

345. ***John Whicker*** *a mariner belonging to the Grey Hound aged 22 years called in & sworn & the above question being put to him answers.*

 No I never heard any such words made use of by ***Carter*** *nor was I ever walking upon the deck and talking with him since the cutter has been in dock.* [Signed with his full signature].

346. *Q & A The same question namely No 339 being put to* ***Robert Johnson*** *a mariner on board the Grey Hound aged 25 years. He answers that he has frequently walked with* ***Carter*** *upon the Grey Hound's deck, but declares upon his oath that he never heard said* ***Carter*** *make use of an such expressions.* [Signed with his mark].

347. *Q & A The above question being put to* ***John Williams****, aged 20 years, one of the Grey Hound's crew, he answers in the negative.* [Signed with full signature].

348. *Q & A The above question being also put to* ***Joseph Wallace*** *aged 21 years another of the said crew, he answers in the negative.* [Signed with his mark].

349. *Q & A* ***William Bolt*** *a mariner of the Grey Hound Cutter aged 22 years, answers to the above question in the negative.* [Signed with his full signature].

350. *Q & A The above question No 339 being put to* ***William Stiles*** *one of the Grey Hounds crew aged 22 years he also answers in the negative.* [Signed with his mark].

351. *Q & A* ***Thomas Duke*** *aged 21 years belonging to the aforesaid cutter answers also in the negative.* [Signed with his mark].

352. *Q & A* ***Charles Cowan*** *a mariner of the Grey Hound aged 33 years answers also in the negative.* [Signed with his mark].

353. *Q & A George Stiles a mariner of the Grey Hound aged 28 years answers also in the negative.* [Signed with his full signature].

354 *Q & A* ***William Murran*** *aged 22 years a mariner belonging to the Grey Hound answers also in the negative.* [Signed with his mark].

355. *Q & A Mr* ***John Carter*** *being again called in and directed to give an answer to that part of* ***Baggs*** *charge in his answer No 308 Wherein he states that said* ***Carter*** *came upon deck to him and asked if he had told* ***Captain Watson*** *what he had heard him say & what it was that he did hear him say for that he did not recollect. Replies that what* ***Baggs*** *has stated in said answer is not true but the following is a true state of the case. 'One morning I was up on deck and* ***Baggs*** *was there & having been informed (but I do not recollect by whom) that he had told* ***Captain Watson*** *something that I should have said to the Captain, I asked him what it was that he had said to the Captain about me & he replied that he was standing upon the deck & heard me say what is recited in his evidence which is now read out to me. And I answered that he never heard me say any such thing, nor any other person overheard me say so. And this upon my oath I do declare to be true.'*

356. *Q & A* ***Stephen Watson*** *the 2nd of the mariners of the Grey Hound Cutter being called in and sworn is required to declare upon his oath whether ever he heard* ***John Carter****, the Mate, make use of the following expression to him or any of the crew as he was walking upon deck since the cutter has been*

> *in dock. - 'I am afraid the old man (meaning the Captain) will sail with us again?' and the person to whom he addressed himself replying 'Why?' and* ***Carter*** *answered 'I do not think he will be turned out this time unless you all keep in one story.'* [**Watson** signed with his full signature though there appears to be no indication of whether he answered in the positive or negative. This is **Stephen Watson** the 16 year old mariner, not the Captain of the same name].

Taken and sworn this 4th of April 1794 being the conclusion of the evidence [signed by **Burrow & Hammond**]

The following day, 5 April, Burrow and **Hammond** concluded a second report to their superiors on **Captain Watson's** charge. The Surveyors General appear to have sat on the fence regarding the truth of the matter and in effect 'passed the buck' regarding judgement to the Naval Board.

They concluded:-

..*That she* [the Spitfire's boat] *and her crew was once in safety and in the tow of the cutter after they were nearly three leagues out of their way, nay directly the contrary way to Studland Bay for they were far so seen and is perfectly clear Why she was not so preserved in safety, whether her loss preceded from harshness, obstinacy, fatigue, or any other cause in* ***Captain Cook*** *and his crew or from the want of having that actual assistance afforded to her by the Greyhound or that superintending care over her when in tow, which the tempest of the night demanded is a matter (All circumstances considered and the whole of the evidence in books A, B, & C now delivered in, weighed and examined) that must be and is respectfully submitted to the wisdom, consideration and judgement of the Honorable Board by* [signed by **Burrow & Hammond**] *Surveyors General Office 5th April 1794.*

END NOTE

Not to be beaten, Watson submitted an additional defence on 7 April, but no record has been found so far of the man's fate, though a press notice in the Hampshire and Sussex Chronicle of 9 September 1803 records that a new Captain, **Captain Wilkinson,** was then in command of the Grey Hound Cutter.

Bibliography

The following list will provide a good source of information for those wishing to know more about the Cook family, though in the light of the historical revelations presented in this book, caution should be exercised with regards to material facts that may simply have been repeated from those presented in earlier works.

Most of the 18th century newspapers and old books listed can be accessed on various archive internet sites, or via the British Library (website – https://www.bl.uk). Some old books have been published as reproduction copies by various publishers. The Australian National Archives Trove portal ((https://trove.nla.gov.au/) is another good source of old newspaper articles and other information.

A Legacy of Mystery – a series of articles in three parts by Alan W Smith of the Captain Cook Society 1996, published in "Cook's Log" the quarterly journal of the Captain Cook Society.

An Historical Account Of The Circumnavigation Of The Globe, And Of The Progress Of Discovery In The Pacific Ocean, From The Age Of Magellan To The Death Of Cook – Harper & Brothers, New York 1837.

Archaeological Investigations at St Andrews the Great, Cambridge; University of Cambridge, Miller J, BA 1992.

Australian Town and Country Journal – 31 May 1890

Biographical Register of Christ's College, II – Peile, J. – Cambridge, 1913. ISBN 978-1107426061

Captain Cook – Walter Besant – Cornell University Library, 2009 ISBN-13: 978-1112190650

Captain Cook: Master of the Seas – Frank McLynn – Yale University Press; 2012. ISBN 978-0300114218

Christ's College in Former Days – Edited by H Rackham M.A. Fellow of the College – University Press, Cambridge 1939. ASIN: B000ZOTEWQ

Evening Post, (New Zealand) – 7 August 1928 & 8 October 1928

Gentleman's Magazine (UK) – February 1794 & May 1863

Hampshire Chronicle – 4 February 1794

Hampshire and Sussex Chronicle – 9 September 1803

Hawaiian Journal of History, vol. 25 (1991)

Ipswich Journal, newspaper (UK) – 25 March 1843

Le Keux's Memorials of Cambridge Vol II – (1842)

Lloyd's Evening News – (London), Wednesday 22 February 1794

Lloyd's Evening Post – 14 October 1793 & Wednesday 12 February 1794

London Morning Post – 3 November 1817

London Public Advertiser or Political Literal Diary – 29 January 1794

The Life of Captain James Cook – Arthur Kitson – (1907)

Malton Messenger newspaper (Yorkshire) – 28 September 1867

Mariner's Mirror, Nov 1962; Mitchell Library, Sydney – N387-06/2.

Morning Chronicle newspaper (UK) – 16 November 1793 & *28* January 1794

Naval Chronicle 1799: (read online – various internet Archives)

New Zealand Evening Post – 7 August & 8 October 1928

Ohinemuri (New Zealand) Regional History Journal – September 2011

Public Advertiser or Political and Literal Diary (London) – 29 January 1794

Reading Mercury newspaper – 3 February 1794

Salisbury Journal newspaper 3 & 17 February 1794

Saturday Journal (UK) 3 February 1794

Star newspaper London, England – Monday 7 October 1793 & 30 January 1794

Steel's original and correct list of the Royal Navy – David Steel 1794 ASIN: B009H0ASG0

St. Andrews church burial registers – Cambridge, U.K

Sun newspaper (London), Saturday 16 November 1793

Sussex manors, advowsons, etc., recorded in the Feet of fines, Henry VIII. to William IV. (1509-1833); ed. by Edwin H.W. Dunkin") – Wentworth Press – ISBN-13: 978-1372413667

Sydney Morning Herald (Australia) – 30 January 1934 & 5 August 1936

The Life of Captain James Cook – Arthur Kitson – Book Jungle. ISBN-13: 978-1438520346

The Life of Captain James Cook – J. C. Beaglehole – Stanford University Press – ISBN 978-0804720090

Times newspaper, Tuesday 28 & 30 January 1794

West Australian newspaper 24 October 1934